ISBN 978-1-330-08200-3
PIBN 10021170

For support please visit www.forgottenbooks.com

1 MONTH OF FREE READING

at

www.ForgottenBooks.com

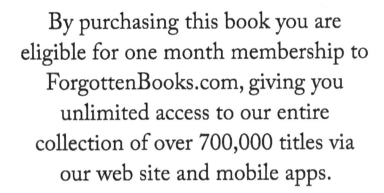

By purchasing this book you are eligible for one month membership to ForgottenBooks.com, giving you unlimited access to our entire collection of over 700,000 titles via our web site and mobile apps.

To claim your free month visit:

www.forgottenbooks.com/free21170

Similar Books Are Available from
www.forgottenbooks.com

The "Westminster" Series

ELECTRIC POWER AND TRACTION

ELECTRIC POWER

AND TRACTION

BY

F. H. DAVIES, A.M.I.E.E.

LONDON

ARCHIBALD CONSTABLE & CO. LTD.

10 ORANGE STREET LEICESTER SQUARE ₁W.C.

1907

BRADBURY, AGNEW, & CO. LD., PRINTERS,
LONDON AND TONBRIDGE.

CONTENTS

CONTENTS.

ELECTRIC POWER AND TRACTION.

INTRODUCTION.

In common with practically every industry Electrical Engineering has become the centre of a wide and increasing circle of allied trades, of which probably the majority are connected in some way or another with its Power and Traction branches. To the members of such trades and callings, to whom some knowledge of applied Electrical Engineering is desirable if not strictly essential, this book is particularly intended to appeal. It may also be useful to the student as it covers with but small exception purely practical ground, and enters to some extent into those commercial considerations which in the long run must overrule all others.

For obvious reasons it is impossible to incorporate in one volume the whole theory and practice of a subject so far reaching as Electric Power and Traction; it therefore becomes necessary at the outset to credit the reader with a certain amount of scientific and technical knowledge: to fix the standard, let us say, such as should naturally have been acquired by observation in the practice of an allied trade and by general reading in an age that is essentially electric.

Not the least of the difficulties which surround the task

of catering for such a circle of readers is that of determining exactly at what point to draw the line ; of knowing how much to tell and how much to leave unsaid. Whatever standard is adopted it goes without saying that a certain percentage of readers must find something of which further explanation would be welcome ; while others who start with a better knowledge of first principles will be able to pass over much as redundant.

At the end of the book will be found a Glossary, which, while pretending in no way to completeness, will, it is believed, serve the purpose for which it is intended, namely, to explain concisely such technical terms and expressions as may reasonably be expected to be outside the sphere of the novice.

CHAPTER I.

To arrive at a clear understanding of the why and wherefore of matters electrical as they exist at the present day, it
will in many cases be necessary to free the mind to a large
extent of preconceived notions; and perhaps in no section
of the subject will this course be more desirable than in
that relating to the generation of Electric Power.

We are now concerned solely with electricity as it is produced commercially, and for this reason, such generators
as the Frictional Machine, the Primary Battery, and the
Thermo-Pile may be left out of the question, attention being
paid exclusively to the dynamo, at present the only machine
capable of generating electricity on commercial lines.

The whole question resolves itself into one of cost and the
survival of the fittest. If, for instance, a primary battery
can be invented capable of generating electricity at a lower
cost and with equal convenience, the dynamo must unquestionably resign its sway. But, as such a revolution in
methods is, so far as we can see at present, extremely unlikely
to take place, alternative systems of generation may with
safety be banished from the mind.

It may be assumed that the fundamental principle of the

dynamo, namely, the manner in which current is generated in its armature, is understood; as it is the elementary process upon which the whole of the electric power industry is grounded the assumption should in this case be justifiable. In like manner, the chief electrical measures such as the Volt, Ampere, Watt, Kilowatt, Board of Trade Unit, etc., and their relation to one another must be taken as within the scope of the reader's knowledge. Assuming this foundation it is possible to pass direct to technical details and examine the manner in which electricity is generated commercially in a power station.

It is not proposed to deal with the mechanical side of the question, and it must suffice to say that practically all forms of prime movers are in use for the operation of dynamos. In the small private power house we often find the Oil Engine; and in some public central stations also this type of prime mover is in use in a form suited to the larger power required. The Gas Engine is making headway, and there are several power houses both in Great Britain and abroad equipped with such engines. Still, in by far the majority of works the steam engine will be found in one of its many forms, ranging from the small old-fashioned horizontal engine to the giant steam turbine, the latest thing in prime movers.

The equipment of a Power Station, the size and class of plant installed, and the manner of its working will vary considerably according to the purpose for which it has been put down, and it is therefore impossible to generalise on the subject. At the least we must consider three types of generating station: (1) such as would be put down to supply the power and lighting requirements of a compact medium sized provincial town ; (2) one suitable for a similar but straggling or more extended district ; (3) the large power company or railway generating station. Between these three there is very little similarity, the requirements in each case being different, and calling for plant and methods of working of a different nature.

In every case the problem before the designer is to generate and distribute the power economically and efficiently, and this is best effected in the first instance of a compact area by the installation of a low pressure continuous current system.

In the early days of electric lighting when the demand for current and the area over which it was supplied were both small, it sufficed to run a simple circuit of two wires from the dynamos, and to generate at a pressure of 50 volts, or 100 volts as a maximum. But as the demand grew, and as the distributing mains were extended, it was found that these latter were becoming inconveniently large and expensive; indeed, it became apparent that economical distribution of electric power could only be carried out in a very small way by such methods owing to the disproportionate cost of the cables.

The solution to the difficulty was found in higher pressure. Electric power in Watts is the product of the Volts and the Amperes; if, therefore, the pressure of supply be doubled, the amperes remaining the same, double the amount of power can be transmitted through a cable without increasing the quantity of copper. This is exactly what was wanted ; but now a formidable difficulty arose, for while an efficient and reliable incandescent electric lamp could be made for work at 100 volts pressure, a 200-volt lamp was in those days never seen outside the walls of the laboratory, being both inefficient and costly and possessing only a very short life. A 200-volt motor could of course be made since it was possible to build dynamos to work at that pressure, but at this period electric motive power was very much in its infancy, and lighting was the main point to consider.

As is frequently the case with epoch making inventions, the means by which this difficulty could be got over was discovered simultaneously by two prominent scientists, viz., Hopkinson in this country and Edison in America. Both proposed a modification termed the Three-wire system of distribution, which, by reason of its simplicity and effectiveness at once sprang into favour, and is, combined with certain improvements, in universal use at the present day.

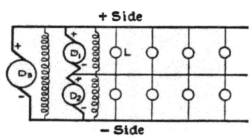

FIG. 1.—DIAGRAM OF THREE-WIRE SYSTEM.

The original idea was as given in diagrammatic form in Fig. 1. Two 100-volt dynamos D 1 D 2, termed after their function " Balancers," are connected in series, so that conjointly they generate current at a pressure of 200 volts. To these are connected in the manner shown the three cables which convey the current to the lamps L. Now it will be obvious that so long as the number of lamps (of equal candle-power) is the same on each side, the current will flow straight through them, and the third or middle wire will be useless and need not be installed. It was foreseen, however, that such an ideal state of affairs could not be counted upon to exist in practice, even if the greatest amount of care were expended on the arrangement of the subsidiary circuits tapped off the main cables. Without the third wire, the result of inequality in the number of lamps on the two sides would be that those which constituted the smaller number would burn too brightly, the pressure being too high, while the other side would suffer from low pressure and consequent bad light. The use of the third wire, however, completely gets over this difficulty. Assume that each of the eight lamps in the diagram requires half an ampere at 100 volts; when they are all switched on a current of two amperes at 200 volts will flow from the two dynamos, and half an ampere at 200 volts will flow through each of the four pairs of lamps, every individual lamp absorbing 100 volts or half the total difference of pressure between the wires + and −

Now assume that one of the lamps on the lower or side of the system is switched off: the condition of exact balance is at once upset and the third wire comes into use, conveying back to D 1 the current taken by the fourth or extra lamp on the + side. There will thus be half an ampere flowing in the third wire, 2 amperes in the + wire, and 1½ amperes in the −. In the same manner, if the conditions be reversed and one of the lamps in the upper or + side be switched off, the middle wire will then convey current from D 2 to the extra lamp on the − side. To summarise, the third wire only carries current to or from the dynamos when the quantity required on one side is in excess of that on the other.

In all three-wire generating stations the main part of the

load is carried by larger generators connected straight across
the outer wires as shown by D 3 in Fig. 1. The balancers
are machines of smaller capacity as well as of lower voltage,
and connections to the mains are so arranged that the
difference in load on the two sides shall never exceed the
capacity of one of these machines.

The system of engine driven balancers described above is

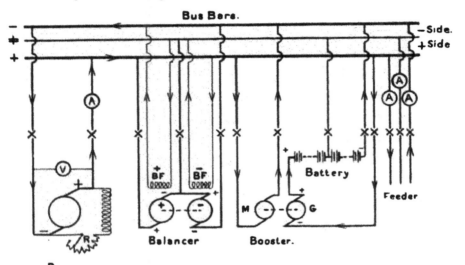

FIG. 2.—DIAGRAM OF CONNECTIONS OF THREE-WIRE DIRECT CURRENT
CENTRAL STATION. THE ARROWS INDICATE THE DIRECTION OF FLOW
OF CURRENT.

in considerable use at the present day, but it does not
represent the most recent practice. Of late years a cheaper
and almost entirely automatic arrangement, which is shown
in diagram form in part of Fig. 2, has been devised for
balancing purposes. This diagram, as a whole, gives the
chief connections of a three-wire continuous current central
station ; and although at first sight somewhat formidable,
it will be found comparatively simple if considered in detail.

On the extreme left are the connections of one of the main dynamos, joined up to the two outer conductors of the three-wire system. For the moment, however, it will be as well to leave this for later consideration, returning to the subject of balancing. To convert the method shown in Fig. 1 to the more modern system of balancing, three radical changes are necessary: firstly, the engines must be taken away; secondly, the two machines must be mechanically coupled together, preferably by a rigid connection of the shafts as indicated by the dotted line in the diagram; and thirdly, the field magnet shunt windings (B F) of the machines must be connected to the opposite side of the system to that to which each machine belongs. This it will be noted is done in Fig. 2, the + balancer field being excited from the — and third wires, while the — balancer shunt field circuit is connected to the third and + wires. When the system happens to be perfectly balanced for the moment, as occasionally occurs, these two machines will run as motors, doing no work beyond revolving their armatures. But should the load be increased, say on the + side of the system, the pressure will naturally fall on that side, causing less current to flow through the field circuit of the — machine, which will have the effect of making it run faster. The two being coupled together, the + machine will be forced up to a speed higher than that which it was running at as a motor, with the result that it will now begin to generate current as a dynamo and so increase the pressure to its right value. It will be seen that with a varying balance the two machines will each be running alternately as a motor and a dynamo. This is possible for the reason that a continuous current motor is simply a dynamo with

its functions reversed.   The dynamo is supplied with power
which it converts into electrical energy ; the motor is
supplied with electrical energy which by reason of the
magnetic fields it sets up in the magnets and armature
causes the latter to rotate and develop mechanical power.

The next series of connections in Fig. 2 are those of the
battery and its attendant booster.   In an installation such
as we are now dealing with, a set of accumulators is an
indispensable adjunct.   It enables the plant to be shut
down at times when the demand is slight, thus saving
wages, and generally adds to the efficiency of the works
by providing load for the machines in the shape of charging current at times when they would otherwise be running
underloaded and consequently inefficiently.   Furthermore,
a battery is often invaluable in the event of a breakdown or
partial disablement of the plant, and its use materially
reduces the pressure fluctuations which are bound to occur
in a station supplying a varying load.

It will scarcely be necessary to go into details of the theory
and construction of secondary cells ; and for the purpose
we have in hand it should suffice to deal only with such
particulars as affect their operation.   The normal voltage
of one element of a secondary battery—that is to say, of one
cell—is two volts : for a 100-volt three-wire supply system,
therefore, at least 50 cells connected in series will be required
on each side.   As, however, at the end of the discharge
the E.M.F. of each cell will sink to about 1·8 volts, it is
customary to provide an additional 5 cells, which, by aid
of a regulating switch, may be cut in or out of circuit as
occasion requires.

When the battery is fully discharged, that is to say when

the E.M.F. of each cell has fallen to 1·8 volts, recharging is performed by passing a current through it from the dynamos in a direction opposing that in which the battery tends to discharge. For this purpose the + pole of the dynamo is joined to the + pole of the battery, the two − poles likewise being connected. If the battery connections in Fig. 2 are followed it will be seen that this is the arrangement. Now, so long as the E.M.F. of the battery is below that of the dynamos the latter will, of course, be able to pass current through the cells. But as the battery becomes charged its E.M.F. rises, until finally it equals that of the dynamos and no current can flow. Of course, if the dynamo pressure be increased, as may easily be done by raising the speed or increasing the excitation of the fields, current will again flow; but this would mean raising the voltage over the entire system, which for obvious reasons would not be permissible. It becomes necessary, therefore, to treat the battery independently, placing it in a separate circuit of higher pressure. This increased E.M.F. may be provided in a number of ways; for instance, a separate dynamo might be used, the battery being entirely disconnected from the supply system. But for several reasons such a course is undesirable, and the usual method is based upon the lines shown in Fig. 2 in which a machine, termed in Americanese a "Booster," is employed. There are many kinds of boosters, differing materially in their arrangement and connections; all, however, are designed and installed for the purpose of varying the pressure in one particular circuit, and the type under consideration is the simplest of the series. It consists of a motor, M, supplied with current from the main circuit, and a generator, G, coupled direct to the motor as indicated by the dotted line. The E.M.F. of this generator will in the case of a 100-volt installation be anything from zero to 60 volts, the range being obtained by varying the excitation current in the usual way by means of an adjustable resistance placed in series with the field winding.

If the connections (which are those for charging) are followed it will be seen that the generator of the booster is in series with the current tapped off the 'bus bars, and by this means a pressure of 260 volts can, if necessary, be applied to the battery. At the commencement of the charge the booster will be cut out of circuit, since the normal generator pressure of 200 volts will suffice to force the necessary charge through. As the pressure of the cells rises the amount of current will drop, and the booster will then be started up and regulated to give the extra pressure necessary to maintain the charging current at its right value. This pressure will have to be steadily increased throughout the period of charge, until at the end it will have reached the maximum of 60 volts obtainable from the machine. In a system of this description the switch (indicated by ✕ in the diagrams) in the cable joining the centre of the battery to the middle wire will be open, and the charging current will go straight through the whole battery. In another booster system of charging two generators are used, one for the half of the battery connected to the + side of the mains, and one for the − half of the battery. This has the advantage of flexibility, since the two halves of the battery may be treated independently, which is often desirable.

In connection with boosters generally it should be understood that there is no radical necessity for them to be composed of a motor and a generator; a few cells of a battery might for instance be used, or the dynamo might be driven by an engine instead of a motor. In either case the arrangement would, in effect, be a booster, since it would "boost" or add to the voltage of the circuit. In common parlance the term when used without any qualification would mean a set such as has been described above.

Before leaving the subject of batteries as installed in a three-wire continuous current station, attention must be drawn to the fact that they are extremely useful as balancers, being often made to serve this purpose. If the load on one side becomes heavier than that on the other, the half of the battery connected to that side can be discharged by the switching in of a few regulating cells; and if necessary, load can be put on the other side to balance up by running up the booster and charging that side of the battery.

CHAPTER II.

Type and Size of Direct Current Generating Plant—Switchboards
Switches and Instruments—Combined Lighting and Traction
Stations — Shunt and Compound Wound Dynamos — Direct
Current Three-Wire Mains — Feeders — Distributors — House
Services.

In a three-wire direct current central station the generating plant will consist of a series of continuous current
dynamos, in accordance with modern practice almost always
of the multipolar type and wound to generate at a pressure
of approximately 400 to 500 volts. Their size may be
anything from 50 k.w. to 2,000 k w. capacity according to
the magnitude of the undertaking.

From the dynamos the current is led by cables to the
switchboard, which is often situated on a gallery so as
to command a good view of the engine room. It occupies
to a great degree the same position of importance as
the bridge of a ship, as at this central point the main operations of the station are controlled by the various instruments
and switches, which when mounted on slate or marble
panels form the switchboard. To make detailed mention
of these would be fruitless, since arrangements vary considerably with the make of board. All low-pressure continuous current boards, however, follow certain definite
lines, being capable of sub-division roughly into five parts,
viz.: Generator, Feeder, Battery, and Balancer panels and

Omnibus Bars. These latter are common to all the panels, and consist of three heavy copper bars to which the dynamos, balancers, and battery are connected through their switch-gear and instruments as shown in the diagram, Fig. 2. The feeder cables that run to various parts of the town are also connected to these bars, which thus link up the street mains with the generators.

Taking the panels in detail, on each of those devoted to the dynamos will be found an ammeter and voltmeter (A and V, Fig. 2) which show respectively the amount of current being generated by the machine and its pressure. In addition, there will be a registering watt-meter to record the number of Board of Trade Units turned out by the machine. The switchgear will consist of magnetically operated automatic switches or circuit-breakers, designed automatically to cut the generator out of circuit should it for some reason become heavily overloaded, or as the other extreme, fail to generate current. Each dynamo panel will also contain an adjustable resistance (R, Fig. 2) whereby the E.M.F. of the dynamo may be varied at will within certain limits.

The feeder panels are generally simple, being fitted only with circuit-breakers, ammeters, voltmeters, and sometimes an arrangement of plugs to allow of easy alteration of the feeder connections to the 'bus bars.

The battery panels will vary largely with the type of booster installed, as the switchgear for this is generally mounted on these panels. A booster of the simple type indicated in Fig. 2 would be provided with a starting switch for the motor (see Chap. V.), a circuit-breaker, an ammeter, a voltmeter, and a regulating resistance inserted in the field circuit for the purpose of varying the speed. The generator side would be fitted with a circuit-breaker, an ammeter, a voltmeter connected across the terminals to show the amount of "boost," and a shunt regulating resistance for controlling the E.M.F. of the machine. Where the discharge of the battery is regulated by the cutting in or out of additional cells, heavy multiple contact switches will be provided for this purpose.

In the case of the motor balancer illustrated in Fig. 2 the switchgear will consist of starting switches, circuit-breakers, and field regulating resistances : there will be two voltmeters, one across each machine, and what are termed polarised ammeters to indicate by the scale to which the needle points whether the machines are motoring or generating, and to what amount in amperes.

To avoid misconception it may with advantage be reiterated that the above description must be taken only as a general indication of direct current switchboard practice. Many arrangements, particularly those relating to boosters and balancers, are much more complicated, and will call for the use of still further switching and measuring apparatus.

So far, we have dealt with the Direct Current Central Station as designed solely for the supply of current for lighting and power purposes. A large proportion of such stations, however, are arranged for the operation of tramways in addition to their primary function, and this calls for some modification of the plant and switchgear. As a tramway load is essentially a rapidly fluctuating one owing to the frequent starts and stops, it is undesirable to use the same generators for supplying the two loads simultaneously, as the pressure fluctuations would have an extremely bad effect on the life of the lamps. It is therefore customary to provide entirely separate feeder circuits for the tramways, with separate feeder panels and 'bus bars. Lighting and power supply on the continuous current system is now usually carried out at a pressure on either side of the three-wire network of from 200 to 240 volts, therefore the main machines, which generate at from 400 to 500 volts, are quite suitable for the traction load which in modern practice is supplied at the latter pressure. When destined for both purposes the generators are invariably compound wound, since this type of field winding is the most suitable for a varying load owing to its compensating effect on the pressure. A sudden increase of the load on a dynamo will naturally cause its pressure to fall, and if the machine be shunt wound the effect will be cumulative, since less current will be passed round the winding and the magnetic field will become weaker. In the opposite manner, a decrease in load causes more current to flow in the shunt, a stronger magnetic field, and increased pressure. Now, if a series coil be added to the field winding, as is the case in compound wound machines, a sudden decrease in the load weakens the field which is partially dependent for its existence on this coil, and the proportion which the two windings, the shunt and the series, contribute to the total result is so arranged that the pressure remains approximately constant within fair limits of variation. Similarly, an increase of load strengthens the field by the agency of the series coil.

In a station of this description the generator panels of the switchboard will each be fitted with what is termed a throw-over switch, constructed so that the machines may be connected either to the lighting or traction 'bus bars but not to both at once. As tramway work is always conducted on the simple two-wire principle no balancers are necessary.

Further details of electric traction will be found in the chapters under that heading. Meanwhile we may next consider the mains system of a three-wire continuous current installation and the means generally which are taken to distribute the current.

By Board of Trade regulation it is required of all supply authorities to maintain their pressure constant within 4 per cent., and to do this at all loads necessitates careful design of the distributing system. The feeder cables, which have been previously mentioned, will be of what is termed the triple concentric type; that is to say, the three conductors (each composed of several strands) are arranged concentrically with layers of insulation between them, being finally covered with a lead sheath. If for laying direct in the ground, a further layer of insulation, a helical steel tape sheath or wire armouring, and an outside covering of braiding will be added. A cable of the former description must be laid either in iron or stoneware pipes or in wood or asphalt troughing in order to protect it from injury.

From the generating station several feeders will be run terminating at convenient points in the town in feeder boxes or pillars, and from this end of the feeders small wires are run back to the works and connected to the pilot voltmeters on the switchboard. By this means, despite the varying load, it is always possible to keep the pressure correct at these points, the station pressure being raised or lowered according to the indications of the pilot voltmeters. Owing to the loss in the cables which increases with the load, the station pressure is always higher than that at the feeding points.

Under no circumstances are the feeders tapped to supply consumers ; this is the function of what are termed distributors, lengths of three-stranded cable which radiate from the ends of the feeders. They are laid in much the same way as the latter, and will of course vary in size according to the demand for current in the particular street they supply. At each consumer a T joint is taken off the wires, generally by aid of a special box designed to facilitate the operation. The service cable enters the building at some convenient position on the street level, and terminates in cut outs or fuse boxes the function of which is to protect the service from any great excess of current due say to a short circuit in the house wiring. The consumer's main fuse is another affair, this being lighter and intended to protect his wiring only. Distributing cables are always sectionalised, that is to say, their continuity is broken at certain intervals by a section box fitted with terminals and removable links. In the ordinary course these latter are inserted and the distributors thus linked up into a network, but in the event of a fault occurring on any one section the links are withdrawn and the effects of the trouble thereby isolated to that particular length of cable.

The quality of the insulation used for underground cables is a highly important matter. In the early days, rubber and impregnated jute were the favourite materials, but these have been almost entirely replaced by paper,

c 2

vulcanised bitumen, and other prepared substances. A system of bare copper strips supported on insulators in a conduit was at one time much in use for feeders, but this has given place to lead-sheathed or armoured cables laid in the manner above mentioned.

CHAPTER III.

IT has been stated that the transmission of electric power over any great distance is only possible economically by the use of high pressure. For consumption within a radius of 2 miles or so from the generating station, current may with advantage be generated and distributed on the low pressure direct current system; but when, as frequently happens, some of the districts to which supply must be given are situated 3 or 4 miles from the works, high pressure transmission to these points becomes imperative.

There are several instances of high pressure direct current transmission in this country and abroad, working at pressures ranging from 1,000 volts to several thousands. Common practice, however, does not turn this way for reasons which render the use of high pressure direct current desirable only in special cases.

For lighting, current cannot well be supplied at a pressure of over 250 volts, since incandescent lamps suitable for higher pressures are not made. Likewise, for power 500 volts is the limit at which it is desirable to supply direct current motors, as above this the pressure is decidedly dangerous to life and the machines are apt to give trouble under unskilled attendance. If, therefore, high pressure direct current is generated at the works and transmitted to the feeding points, it becomes necessary there to provide transforming apparatus to reduce the supply to a pressure not exceeding 250 volts, or 500 volts between the + and − cables of a three-wire system. At the present time the only commercial way to transform direct current from one pressure to another is to make use of a machine termed a motor-generator or motor-transformer. This consists simply of a direct current motor designed to work at the high pressure delivered at the feeding points, coupled mechanically to and driving a direct current dynamo wound to generate at some suitable low pressure of supply and connected to the distributing mains.

It is thus necessary to provide a number of substations situated at various points of the area, each containing one or more motor-generators. Engineeringly quite possible, such a scheme suffers from one disadvantage ; it is comparatively speaking inefficient. No machine, dynamo or motor, possesses at the present time an efficiency of over 95 per cent. at full load, and at light loads it drops considerably below this figure. It follows, therefore, that under the most favourable conditions a loss which is entirely absent in the low pressure system is constantly incurred, particularly at the period of light load which it must be remembered occupies by far the larger portion of the day with the ordinary lighting and power undertaking. By careful design and management much may be done to alleviate this inefficiency, but it must always be present to some extent. In addition, with a system employing motor-generators, a certain amount of attendance is necessary, and the capital cost is increased by the buildings which must be provided for their reception.

Where high pressure transmission to the feeding points in the town is imperative the method almost universally adopted is that known as the Single Phase Alternating Current system, as this affords the best means for efficient and convenient transformation of the high pressure current. Before proceeding further into details some definition of the expression Single Phase Alternating Current may be desirable.

The rotation of the armature coils of a dynamo past magnetic poles of opposite sign gives rise to currents of electricity in the coils, alternating in direction according to the polarity of the magnetic field through which they are passed. In the case of the direct current machine these are rectified by the commutator; but in the alternating current generator this piece of apparatus is of course absent. The current generated by such a machine is therefore "alternating," and it is termed single phase to distinguish it from the electric power delivered by other classes of alternators designed to generate two or three separate alternating currents simultaneously and named Two-Phase and Three-Phase machines respectively, or as a class, Polyphase generators. These latter will be dealt with in their turn; at present it is only necessary to consider the simple alternating current generated by the single-phase alternator.

In one respect the alternator has an advantage over the

direct current machine; it is capable of generating at very high pressure without risk of breakdown, whereas the latter always possesses a weak point in the commutator. Usually the armature of an alternator constitutes the fixed portion while the field magnet rotates, an arrangement that does away with many disadvantages from which the direct current machine suffers and allows of a better mechanical structure. Practically speaking, it is impossible to build a direct current dynamo with a fixed armature and rotating field, the commutator and collecting gear generally forming the difficulty.

The main reason for the adoption of alternating current for high tension transmission in preference to direct current lies in the ease and efficiency with which the former can be transformed in pressure. In place of the cumbersome and inefficient running machinery required by the latter all that is necessary is a stationary piece of apparatus termed a transformer. In operation it is akin to the induction coil, but the alternate making and breaking of the primary current is of course unnecessary since that current is alternating. Referring to Fig. 4, in which a transformer, T, is shown in diagram, and also to Fig. 3, its construction and action are as follows:—On or through an iron core built up of laminations in the form of very thin soft sheets are wound two separate coils, the number of turns in each being in proportion to the ratio of the two voltages, that at which current is transmitted, and that at which it is desired to distribute. For instance, if it be wished to transform from 2,000 volts to 100 volts, the low pressure or in this case secondary coil will have only one-twentieth of the number of turns in the high pressure or primary.

FIG. 3.—ALTERNATING CURRENT TRANSFORMER WITH CASE REMOVED,
SHOWING CORE AND COILS.

The former winding will be of comparatively thin wire, since the pressure of supply is high, and the current therefore small compared with that in the secondary. The turns of the latter will be much thicker in order to carry the comparatively large number of amperes. Assuming such a transformer to possess an efficiency of 100 per cent., if 10 amperes at a pressure of 2,000 volts be flowing in the primary winding the power generated in the secondary will be 200 amperes at 100 volts, that is to say, 20 k.w. in each case. As a matter of fact the efficiency under the best conditions, viz., at full load, will be about 98 per cent., and the secondary output will therefore be lower than the above.

So much for the construction; the action depends upon the principle that relative motion in the right direction between a magnetic field and an electrical conductor results in the generation of electricity in the latter. In the dynamo the motion is obtained mechanically by the rotation of the armature conductors or the field magnets, but in the alternating current transformer it is the magnetic field that moves in the iron core. The current flowing in the primary is constantly reversing its direction; at a certain point of time it is at zero; it then begins to grow larger and larger, with the consequence that the magnetic field it generates in the iron core increases in strength, the lines of force cutting through the secondary winding as they come into being. This, obviously, will result in the setting up of an E.M.F. in the secondary the direction of which will change with the alternating polarity of the magnetic field set up by the primary current. The current thus induced in the secondary will alternate with exactly the same frequency as that in the primary for reasons which will be obvious.

The self-regulating properties of a transformer are both interesting and important. At first sight it would appear that the primary winding, having a fixed resistance, and being supplied at a fixed E.M.F., would always have the same current flowing through it irrespective of the amount being generated by the secondary. If this were so the transformer would be an extremely inefficient piece of apparatus at light load, but such is not the case. The resistance, of course, does remain the same, but when dealing with alternating currents there is another factor to consider, viz., what is termed "Impedance" or "Self-Induction." An alternating current flowing through a coil sets up an alternating magnetic field, which we have seen above induces electricity in any adjacent coil placed so that the field passes through it. But the field generated besides passing through the secondary coil also passes through the winding that is generating it, that is to say, the primary, and in doing so it must obviously set up an E.M.F. in this coil for the same reason that it does in the secondary. This E.M.F. opposes that which is applied to the primary, and therefore, when no current is being taken from the secondary, it dams back that in the primary and keeps it at a very small value. But the moment load is put on the transformer a current begins to flow in the secondary coil, and a still further magnetic effect is produced in the iron core by that secondary current, with the result that the self-induction or impedance of the primary winding is partly neutralised and more current is allowed to flow. The greater the current in the secondary winding the more pronounced this neutralising effect will become, and the two currents will increase with a rising load and decrease with a falling one practically in proportion. This action is decidedly complex and difficult to follow. It is only outlined above, other and more complicated considerations which influence the effect being omitted because their inclusion could only result in the bewilderment of the novice. For the technical student a very clear explanation of the action of a trans-

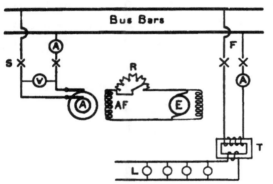

former is given in Dr. Silvanus P. Thompson's book, "Dynamo Electric Machinery"; for the lay reader it should suffice to know that owing to certain magnetic effects in the core the current in the primary auto-

FIG. 4.—DIAGRAM OF SIMPLE SINGLE PHASE ALTERNATING CURRENT SYSTEM.

matically increases and decreases in proportion to that in the secondary, otherwise the "load" on the transformer. Conversion by this means may of course take place in either direction, that is to say, a transformer may be used for converting alternatively to high or low pressure. In either case the coil which is supplied with current is termed the primary, the secondary coil being always the one in which the current is generated.

Fig. 4 is a diagram of the arrangement of a simple Single Phase Alternating Current central station and system. High pressure current is led from the alternator, A, through the usual switchgear and instruments to the 'bus bars, along which it flows to the feeder, F, having at its end the transformer, generally placed in a small chamber under the street. This converts the high pressure current to low pressure as above, and the latter is supplied direct to the distributors at a pressure of from 100 to 240 volts as the case may be. The three-wire system is often used for alternating current distribution in order to save expense in the mains, in which case two transformers are employed, one on each side of the system.

It has been mentioned that the field magnets of an alternator require excitation from some independent direct current source. This is provided as shown in Fig. 4 by a small shunt wound direct current machine termed an exciter, driven by its own engine or off the shaft of the alternator. AF is the alternator field circuit, the current through which is controlled by means of the regulating resistance, R.

The most usual pressure at which current is generated in a single phase alternating current station is 2,000 volts, and the frequency, that is to say, the number of complete periods per second, is generally 50, 60, or 100. A complete period it should be noted comprises two alternations of the current, and a periodicity or frequency of 50 cycles per second therefore means that the current alternates in direction 100 times a second.

The plant in an alternating current station of this type will consist of a number of alternators of sizes varying according to the conditions. The exciters are often supplemented by a battery of accumulators, both from considerations of reliability and efficiency, as the running of a separately driven exciter capable of generating sufficient current for several machines is uneconomical during the period when one alternator is sufficient to deal with the load. There is of course no battery for the main supply since accumulators are continuous current generators and cannot be used with alternating current. Cases exist where a battery in conjunction with a direct current motor driving an alternating current generator is used for supplying the light load when it is desirable to shut down the engine driven plant in the interests of economy, but such an arrangement is very inefficient and suitable only for the smallest undertakings.

The switchboard of an alternating current station is very different in appearance from that of a direct current, and simpler both in construction and working. Each generator will be connected to the 'bus bars by a special form of switch, usually designed to break the circuit in a bath of oil so as to suppress arcing at the contacts. Various makes differ materially in design, but all are or should be arranged with a view to reducing the amount of risk to the operator to a minimum by the covering up or placing out of reach of all parts connected to the high pressure.

In connection with the generators, one piece of apparatus will be found on the alternating current switchboard which is entirely absent in a direct current central station, viz., the instruments used for synchronising. In direct current working when it becomes necessary through increasing load to connect another generator to the 'bus bars, all that has to be done is to see that its pressure is exactly equal to that of the running machines. In an alternating current plant, besides this, it is imperative that the incoming machine should be in exact step with those which are running; that is to say, the frequency of the reversals must be the same, and the E.M.F. it is generating must have the same direction. At first sight this would appear to be a state of affairs by no means easy to bring about or to ascertain; but although the operation itself calls both for skill and experience, the means by which it is made possible are comparatively simple. In a much used system, a small alternating current transformer with two primary windings and one secondary is employed. One of the former windings is connected to the 'bus bars, and is designed to generate in the secondary coil exactly half the voltage necessary to light a lamp connected to it, or preferably, to show a reading on a voltmeter of half a certain prearranged value. The second primary coil is joined up to the terminals of the incoming alternator, and as it consists of an equal number of turns to the first coil, it has therefore an equal generating influence on the secondary. Now, if the currents in these two coils are completely out of phase or step, that is to say, are flowing all the time in opposite directions, the magnetic effects they have on the iron core of the transformer will exactly neutralise one another, and there will be no magnetic field, and consequently, no current in the secondary or light from the lamp. On the other hand, if the primary currents are identical in every respect, the magnetic effect they have on the core will be cumulative; full pressure will be generated in the secondary, and the lamp will light or the voltmeter indicate the correct reading. In the process of connecting a fresh alternator to the 'bus bars this desired effect is brought about by gradual alteration of the speed of the machine, the frequency of the reversals of E.M.F. being dependent upon this. At the critical moment when the lamp or voltmeter shows that the secondary pressure is at its highest, the main switch of the machine is quickly closed and the operation is completed. It is of course quite possible for the incoming machine to possess exactly the same frequency of alternations as those running and yet for the direction of flow or pressure to be exactly opposite. In such a case the speed is slightly altered and then brought back to the correct value, the chances being that the direction as well as the frequency will be right the second time.

CHAPTER IV.

THE GENERATION AND DISTRIBUTION OF POWER.

The Relation of Output to the Cost of Production—Economy of Power Schemes—Polyphase Alternating Currents—Generation and Distribution of Polyphase Currents—The Three-Phase System—Extra High Tension Transmission—Overhead Lines— Comparative Cost of Overhead and Underground Transmission Mains—Wood and Steel Poles for Overhead Lines—Insulators— Lightning Arresters—100,000-volt Underground Transmission Cable.

THE systems of generation and distribution described in the foregoing chapters are those which, with modifications, have been the rule for several years, and of both there are at present a large number in operation in this country. They have served their purpose well and will continue to suffice in many cases for years to come; but in the light of recent developments they cannot be regarded as the permanent solution to the problem of electrical power distribution. It is as true of electricity supply as it is of other forms of trading, that the larger the concern and the greater its output, the lower will be the cost of production, that is, always supposing that the management is efficient and knows its business; and for this reason the electric power of the future will unquestionably be generated solely at large power stations feeding considerable areas. There are numerous schemes of this type at work on the Continent and in America, and at home The Clyde Valley Electric Power Company, The County of Durham Electric

Power Company, The Lancashire Electric Power Company, The Newcastle-upon-Tyne Electric Supply Company, The Scottish Central Electric Power Company, The South Wales Electrical Power Distribution Company, The Yorkshire Electric Power Company, and others, are all giving a supply of current over extensive districts at rates which would have been impossible if the supply were generated at small local stations.

It will be of interest to examine the reasons for this economy in the cost of production. They may be divided into two classes : those which affect the capital cost of the undertaking and those which reduce the running expenses.

Power transmission schemes of this description are always designed for high pressure working to enable them to supply over a large area, and this gives the promoters a much freer hand in the selection of a site for the generating station, allowing them the choice of several suitable positions. Other things being equal, the cheapest is of course selected, and a saving—often considerable—is thus effected in capital outlay on land. A concern such as a municipality or company which supplies only a comparatively small area is obviously handicapped in this direction, and has usually to pay much more in proportion for a site, possibly of a less convenient nature. A second and very important capital economy is made in the cost of plant per unit of output. A generating set of 2,000 k.w. capacity does not cost twice as much as a 1,000 k.w. set; and further, it does not take up twice the room, which means a saving in buildings and lands per unit of output. These two points apply practically to all details of central station apparatus, and the advantage in capital cost therefore lies decidedly with the large power station. Turning now to the cost of production. The large generating sets employed in such schemes are more efficient than the comparatively small ones installed in a local generating station, and the same statement applies to most of the auxiliary apparatus such as exciters, pumps, etc. The cost of fuel and its handling and the cost of water may be kept at a low figure by selecting a site for the works where such can be obtained at the cheapest rate, and rents, rates, and taxes will also be proportionally much lower than when the station is situated in a town. A further large economy is effected in administrative expenses, and in short, all along the line and in every detail the cost of production will be lower the larger the undertaking.

It has been mentioned that the methods employed in power distribution on a large scale are very different to those which have so far been discussed. In the first place high pressure polyphase alternating currents are used, being better suited for the operation of motors than single-phase current and also requiring less copper in the mains for the conduction at a similar voltage of a given power. If the armature of an alternator be wound with two distinct sets of coils placed alternately round its circumference, it follows that the current generated in what we may regard as the second set in the direction of rotation will lag in its rise, and fall behind the current in the first set. Similarly, if a third set be added, the current in this will lag behind that in the other sets. By the proper spacing of the armature coils with regard to the magnet poles it is arranged in the case of a two-phase machine that the amount of lag shall be exactly one quarter period or

90° of the complete phase; that is to say, the maximum voltage of B phase—the lagging one—will occur at the moment when A phase is at zero preparatory to reversing its direction of flow. This is shown diagrammatically in

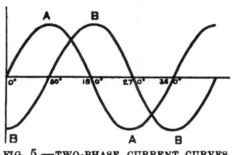

FIG. 5.—TWO-PHASE CURRENT CURVES.

Fig. 5. The horizontal is the time line and is marked in degrees, the period being complete at 360°. The vertical line is the E.M.F. scale. In an alternator constructed to generate three-phase current the sets of coils and the poles are relatively spaced so as to cause the angular difference between the three currents to be 120°, or one-third of a period. This is illustrated in Fig. 6. In either the two-phase or the three-phase system the circuits

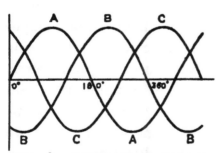

FIG. 6.—THREE-PHASE CURRENT CURVES.

may be kept quite separate throughout; necessitating in the former the connection of four distributing cables to the generator, and in the latter six. In the early days, when polyphase machines were employed, this was done, each circuit being used as an entirely separate single-phase one, and of course, from the point of view of economy in transmission mains there was no gain over single-phase working. It was soon found, however, that four and six wires respectively were not necessary, and that the same power at a similar voltage could be economically transmitted with three wires in either case with a large saving in copper. The majority, and the most important of British power schemes are working on the three-phase system, which has proved to be the most efficient and convenient for the average case. In the interests of brevity the two-phase system may therefore be passed over, attention being confined to the more popular method of the two. It should be stated, however, that much which may be said of the three-phase system regarding its utility for power purposes applies equally to the two-phase. For instance, polyphase motors, irrespective of the system (two or three-phase) upon which they are designed to work, are in many respects superior

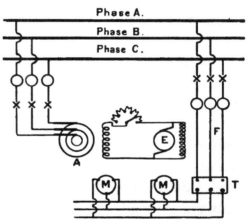

FIG. 7.—DIAGRAM OF THREE-PHASE ALTERNATING CURRENT SYSTEM.

to the single-phase motor; a point which will be more fully dealt with in the section devoted to motors.

Presuming it is now clearly understood that a three-phase current consists of three alternating currents of similar frequency and voltage but out of step with each other to the extent of a third of a period, and that such a current is generated by one alternator having three sets of coils in its armature, and is distributed by three separate wires, we may pass to the consideration of a power scheme designed to work on these lines. Fig. 7 is a diagrammatic representation of such in its simplest form. The alternator A is supplied with current for its field magnet by the exciter E. From the machine, which generates three-phase current at, say, 11,000 volts and 25 periods, three cables are taken to the three bus-bars through instruments and special switchgear designed for use at the extra high pressure. The feeders F are connected to these, and by them the 11,000-volt current is transmitted underground or overhead, as the case may be, to transforming substations situated at suitable points in the area. Here it is transformed down by three-phase transformers to 440 volts or thereabout for distribution to the three-phrase motors M connected to the mains. Where there is a lighting load as well as power the lamps are supplied at the usual pressure of 200 to 240 volts.

All large modern power stations are very similar in their equipment, which, although perhaps intricate in detail, forms as a whole a simple combination. In Chapter XX. the Lot's Road, Chelsea, power station of the Underground Electric Railways of London, Limited, is described in detail so far as the generating plant is concerned, and although larger by far than the great majority of power distributing companies' works, the Chelsea station is designed upon very similar lines, and will serve well as an example. Leaving the details of the power house unconsidered for the present we may now pass direct to the transmitting and distributing system.

With electricity at a pressure of 11,000 volts or so the problem of transmission is by no means a light one. Firstly, the safety of the public, and to a certain extent that of the employees, has to be considered, and then follow numerous economic and technical questions bearing upon the cost of the line and its immunity from breakdown.

From the point of view of the Power Company, whose business it is to earn dividends, a transmission system employing bare wires carried overhead on poles is by far preferable, since it is the cheapest to instal and to maintain. On the other hand, it is, or at any rate is credited with being, dangerous to the public, and, but in exceptional cases, is not permitted in this country for that reason. Abroad a more open view of the question is taken, and we find numerous pole lines transmitting current at as much as 20,000 volts pressure through sparsely populated districts, and in some cases even through towns. In America there are several instances of 30,000-volt overhead transmission lines, and a few are working at 40,000 volts and even higher pressures. The percentage difference in cost between the overhead line and the underground cable is very marked at all pressures, and it increases rapidly as the voltage is raised, owing greatly to the expense incurred in the insulation of the latter. At the comparatively moderate pressure of 6,000 volts the former system under exactly equal conditions of transmission efficiency will cost for an equal power approximately half as much as the latter; and at 12,000 volts the proportion in some cases may even drop to one-third. Much, of course, depends upon the difficulties that are met with in obtaining wayleaves for the wires where they cross private property, and such-like troubles which may cause wide deviations from the straight line run of the wires. But experience abroad has proved that although for these reasons the cost per mile may vary in different instances over a comparatively wide range, overhead transmission from the economic point of view is by far to be preferred. The introduction of such methods to Great Britian has met

with consistent and powerful opposition from the start, and
there is no doubt that the distribution of electric power
generally and the many industries which would benefit
from a cheap supply have been much hampered thereby.
There are, however, indications at least of a change of
feeling in this important matter, and judging from these,
it is permissible to anticipate in the near future ameliorated
conditions for the power supplier in this respect.

The methods employed in the construction of overhead
transmission lines have greatly altered for the better during
the last few years.   This indeed has been a necessary outcome of the increase in pressures.   In the early American
overhead lines it was customary to use rough, unseasoned
poles fitted with insulators of a primitive and unsatisfactory
type, and the result was a system which, if cheap in first
cost, was expensive and troublesome in upkeep.   As such
lines are installed now, the poles are carefully selected and
impregnated with creosote to resist decay.   Various woods
are used, but pine, chestnut, or cedar are found to be the
most suitable.   The poles are fitted with cross-arms after
the manner of a telegraph post, and on these are fixed the
insulators which carry the conductors.   The type of insulator is a matter of importance where the pressure is high,
as they are electrically and mechanically the weakest part
of the system.   Glass and porcelain are both used, and
excepting that porcelain is stronger, and that glass is
cheaper and has a somewhat greater dielectric strength,
there is for practical purposes little to choose between
them.   An essential feature with all insulators is that the
path over which surface leakage to the pole can take place
should be as long as possible, and for this reason they are

constructed with what are termed petticoats, the insides of which remain comparatively dry in the heaviest storm. A type used for the 60,000-volt transmission line of the Missouri River Power Company is shown in section in Fig. 8. The wooden pin is screwed into the body of the insulator, and for part of its length it is covered with a glass sleeve to keep it dry. The wire runs in the groove on the top, being attached to the insulator by a wire binding round its waist. Numerous other shapes are used, but the object is in every case the same, namely, to interpose a long dry insulating surface between the wire and the pole.

FIG. 8.—SECTION OF LINE INSULATOR.

Not the least important detail of the progress which has of late years been made in overhead line practice lies in the substitution of steel lattice poles for those of wood. In Italy especially there are some very fine examples of lattice poles constructed to carry a large number of wires; and transmission lines of any importance will for the future in all probability be erected on this system.

In the practical working of an overhead transmission one of the chief troubles is that occasioned by lightning. In this country, given proper precautions, there would be little to fear in this respect, but abroad, interruptions of supply owing to the line being struck are not uncommon. The remedy lies in the installation of lightning arresters at approximately quarter-mile intervals along the line. These take many forms, but in principle they are the same in that they provide a path for the discharge to earth without permitting the line current to follow in its wake by maintaining the arc formed by the lightning. One much-used type consists of a number of cylinders made of a metal which does not readily maintain an electric arc. These are spaced about a sixty-fourth of an inch apart, and the two end cylinders are connected respectively to the line and the earth. When a discharge takes place the lightning, being of very high potential, leaps the gaps and so find its way to earth, but in doing so it forms an arc between each of the cylinders, and there is of course a tendency on the part of the line current to follow and keep up the arc. This, however, is hindred to a great extent by the nature of the metal of which the cylinders are composed and their cooling action on the arc, and the tendency is further reduced by a resistance inserted in series with the line and its connection to the arrester. The consequence is that, although an arc is formed, it dies out immediately the lightning discharge has passed to earth. Such arresters are also employed for the purpose of relieving the line of any undue electrical stresses set up in it. If the current be suddenly switched on or off a high potential alternating current circuit the pressure for the moment rises to a considerably higher value than the normal, with the result that the insulation may fail, either by puncture if a cable, or by the breaking down of an insulator if the transmission be overhead. The causes of this effect are complicated and need not be entered into here ; but an arrester is the remedy since it allows the excessive pressure to escape to earth in the same manner as a lightning stroke.

Distribution through towns is carried out largely by overhead wires abroad; but in this country, although the high pressure current might be transmitted on this system, distribution to the ordinary consumer is always likely to be effected by underground cables. In the case of very large consumers the high pressure current is often taken right on to the premises and transformed *in situ*, which, of course, saves the Company the cost of low pressure cables and is therefore desirable. The low pressure distributing mains of a Power Company call for no particular remark as they will not differ from others, but cables suited for the transmission of extra high potential currents are of necessity of a very special nature. For long it was considered that 50,000-volts would be about the limit of pressure for which an underground cable could be designed for reliable working; but at the Exhibition recently held at Milan, a 100,000-volt cable was shown by the Pirelli Company of that town. The conductor is composed of ninteeen strands, having a total sectional area of 162 sq. mm. This is covered with a lead sheath of 18 mm. external diameter, the object being to obtain a smooth surface on the conductor and so to reduce the electrostatic strain. Next to this comes a layer of rubber 2·5 mm. thick, and then follow two further layers of rubber 2·3 mm. and 4·5 mm. thick respectively. The next covering consists of impregnated paper 5·2 mm. thick, and the cable is finished with hemp and lead sheathing. The total thickness of the insulation is 14.5 mm., and the outside diameter of the cable is $2\frac{5}{16}''$. The results obtained with this cable, which has been actually tested up to 150,000 volts, go to show that transmission by this means is electrically quite possible at even higher voltages than those at present used on overhead lines. The trouble, however, does not lie in this: as stated, it is purely a matter of expense, and upon this score the overhead system is decidedly preferable.

CHAPTER V.

THE ELECTRIC MOTOR.

In previous chapters the electric motor has to a great extent been taken for granted ; it now remains to go into this very broad subject more fully. Its early history and the phases through which it has passed will be found both interesting and instructive. They may be read of in many places, and followed step by step with advantage. It is our purpose, however, to deal more with the electric motor as it exists in its present day form, and to examine the characteristics which collectively give to it the premier place among generators of mechanical power.

Motors may be divided into three main classes according to the system of supply upon which they are designed to work. These are, direct current, single-phase alternating current, and polyphase alternating current. With regard to the first, it has been stated that the direct current motor is simply a dynamo with its functions reversed. The current supplied to the field magnet and the armature sets up magnetic fields in each, which, acting upon one another, produce a constant torque or turning moment in the armature.

In addition, the rotation of the armature of a direct current motor in the field provided by its magnets sets up an E.M.F. in the armature for the same reason as it does in a dynamo, namely that the conductors in their rotation cut the magnetic lines of force. This is a very important point that should be clearly understood if the electric motor and its characteristics are to be fully comprehended. When the motor was first known some difficulty was experienced in explaining the fact that although a considerable amount of current could be passed through the armature when stationary, if it were allowed to run up to full speed unloaded this dropped off uniformly as the speed increased, falling to a comparatively insignificant quantity at top speed. Among the theories put forward to explain the action was the hypothesis that the resistance of the armature conductors was in some incomprehensible way increased by their rotation. But this and similar theories were obviously weak and untenable, and before long the correct reason was discovered, namely, that the motor by the rotation of its armature generated an E.M.F. which increased with the speed, and as this E.M.F. was always contrary in direction to that of the supply the higher the speed became the more the current flowing through the armature was cut down. It was further seen that the back E.M.F. as it is termed must always be lower than that of the supply, and lower by an amount just sufficient to allow the requisite quantity of power to pass to rotate the armature against its load. This action is comparable to that of an alternating current transformer, which, as explained, automatically regulates the primary current in proportion to the load on the secondary by the generation of a back E.M.F. in the former winding.

In each case the greater the load the lower the back **E.M.F.** and the larger becomes the current flow.

The suitability of the motor for certain purposes will depend upon the means provided for the excitation of its fields, and it will be necessary to consider each type of magnet winding with the effect it has upon the operation of the motor. As with the direct current dynamo, a motor may be either shunt, series, or compound wound. The former system is the most general, shunt wound motors being suitable for the majority of industrial purposes, but series and compound windings are necessary to meet certain cases. The shunt wound motor when supplied at constant pressure will run approximately at constant speed at any load within its capacity, and its speed will not vary much with a sudden moderate increase or decrease of load. It possesses a good starting torque, and this combined with its constant speed makes it particularly suitable for such work as printing where full load is thrown on the motor at once and where regular speed is a prime necessity. For the latter reason it is also eminently suitable for textile machinery, wood working plant, and similar cases. The speed of such a motor is varied by a regulating resistance connected in its shunt winding. When resistance is inserted, the current in the coils and consequently the intensity of the magnetic field becomes weaker, with the result that to generate the necessary back E.M.F.—which it tends automatically to do—the motor must run faster. On the other hand, if the current in the shunt winding and consequently the magnetic field be made stronger by the cutting out of the variable resistance, the motor will run slower and still generate the same back E.M.F.

The compound wound motor is excited by a series as well as a shunt coil. They are wound so as to oppose each other in effect, with the result that this type of motor besides possessing a close approximation to constant speed and being able to exert a considerable starting effort is capable of dealing with a very heavy overload with very little reduction of speed. It is about equal to the shunt motor in constancy of speed at various loads and in starting torque; its special virtue lies in the way it responds to momentary widely varying loads such as occur in punching, slotting, shearing machines, etc. Assume that a compound wound motor is running light and that a very heavy load is suddenly thrown on it. If there were only a shunt winding this would slow the motor up appreciably for a few seconds, but with the compound machine the series winding comes into play. Owing to the great load the current in this is increased, with the result that the field which is mainly dependent upon the shunt coil is considerably weakened, the magnetic effects of the two coils being opposite. As was pointed out, a weakened field gives rise to a higher armature speed, and the compound wound motor therefore responds much quicker to a sudden and heavy increase in load than the shunt. The opposite effect is also true. A heavy load quickly thrown off a compound wound motor, owing to the weakened action of the series demagnetising coil, and the consequent stronger field, does not cause the machine to race. As with the shunt motor, variation of speed is attained by a regulating resistance connected in the shunt circuit.

The series wound motor is used principally for traction work, and in connection with the driving of such machines as heavy cranes which call for an exceptionally powerful starting effort. This is one of the chief characteristics of the series motor, and it is superior to both the shunt and the compound motor in this respect. The comparatively large current taken by all motors when starting under a heavy load naturally gives rise in the series machine to a very strong field, with the result that the starting effort is considerable. It is much greater than that of the shunt motor, the excitation of which remains approximately constant irrespective of the current in the armature. On the other hand, the speed of the series wound motor varies largely with the load, for the reason that its excitation, being dependent entirely upon the current in the armature, is variable. With a heavy load it slows up as the field becomes stronger, and with a light load it races for the opposite reason. Owing to this latter effect a series wound motor must never be allowed to run unloaded, as the speed would increase to an extent sufficient to wreck the armature and commutator by centrifugal force. A traction motor can of course never be entirely unloaded, and where series machines are used for crane work a brake of some sort is always provided.

We have now investigated the leading characteristics of the three classes of direct current motors. The subject, necessarily, has not been gone into fully in all its details, but enough has been said to give a good general idea of the why and wherefore of each class. Methods of control now claim attention, that is so far as they apply to the ordinary industrial motor, traction work being left for a later chapter. If a direct current motor of any description were suddenly connected to the mains when stationary, a large current

would flow in the armature, since there would be no back
E.M.F. to oppose it and the ohmic resistance of the coils is
very low. This would not only seriously injure the insulation of the conductors by overheating, but it would also put
a severe strain on every part of the machine, cause injurious
sparking at the brushes, and by reason of the excessive

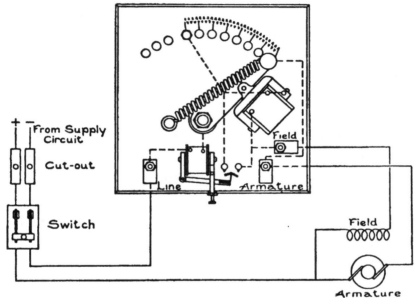

FIG. 9.—DIRECT CURRENT MOTOR STARTER AND DIAGRAM OF
CONNECTIONS.

current, reduce the pressure of the supply temporarily to a
very low value. It is therefore necessary with the motor
as with most other machines or engines to start slowly.
This is done by the interposition of resistance in series
with the armature, and this resistance together with its
multiple contact switch forms what is termed a starting
switch. A modern type of motor starter is shown in Fig. 9.
On the face plate, which is usually of slate, are mounted a

E.P.   E number of brass contacts connected to certain points of a resistance composed of wire or strips fixed at the back of the face plate. The movement of the switch arm across the contacts cuts the resistance out step by step, until on the last stop on the right there is no resistance in circuit with the armature. It will be noticed that there are two coils fixed to the front of the face plate ; that on the right together with its iron core forms a horseshoe electro magnet, and the winding is connected in series with the shunt of the motor. On the switch arm is a piece of iron or keeper, and when the motor is running and the resistance is all cut out the magnet by aid of this keeper holds the switch arm over to the extreme right against the action of a coiled spring which tends to keep it in the " off " position. Now if the supply fails or the shunt circuit of the motor be broken by some mischance, the coil loses its magnetism, and the arm flies back to the left cutting current completely off the armature. This arrangement is termed a " no load release," and is essential to all motor starters. The sudden resumption of the supply after a failure would, if the switch remained on the last contact, result in a bad short circuit and damage the motor for the reasons given above. The thick wire coil at the bottom is termed an " overload release," and its function is automatically to cut the motor out of circuit if the load becomes excessive. The whole of the current taken by the machine passes through this coil, which is wound on an iron core fitted with pole pieces. The keeper is hinged to one of these, and in its normal position lies as shown, resting on an adjustable stop. At its end it is fitted with pieces of spring copper, over which on the face plate are two contact stops connected electrically to the no-load release coil. Now, in the event of the load on the motor becoming excessive the hinged keeper is drawn up by the increased magnetic field generated by the excessive current flowing through the coil and armature, and the copper contact piece makes electrical connection between the two stops above mentioned. As the resistance of this path is very low compared with that of the no-load release coil the current is diverted or shunted from the latter, with the result that the field of the magnet dies away and releases the switch arm which flies back to the "off" position, cutting current entirely off the motor. By these two simple devices a motor is automatically protected against all ordinary mischances. The overload release is adjustable by means of the screw stop on which the keeper rests. The nearer the latter is screwed up towards the pole piece the less will be the current required to attract it, and *vice versâ*. An overload release is not fitted to all starters either on the score of cheapness or because an ordinary and separate circuit breaker is included in the equipment. These take many different forms according to the make, the pressure of the supply, and the amount of current they are intended to break. They are mostly arranged with a coil through which the whole current passes, and the magnetic action of this releases a catch and permits the switch to open quickly under the pull of a powerful spring. The breaking of a heavy current even at a fairly low voltage entails an arc and burning of the contacts unless means are provided to avoid it, and to this end two systems are employed. The first is termed the magnetic blow out and is dependent upon the action of a magnetic field on the electric arc. The latter is itself magnetic, since it is carrying current, and it was found some time back that a powerful magnetic field placed close to an arc would repel the latter, causing it to lengthen and finally break. In the magnetic blow out circuit breaker two sets of contacts in parallel are provided. The one to open first under the action of the spring is the main contact, which in the ordinary way carries most of the current and must not be injured. The one that opens second is the auxiliary contact, and the breaking of this is arranged to take place in a powerful magnetic field produced by the current passing through the switch. The arc occurs, therefore, only on the second contact, which is capable of dealing effectively with it, and no harm is inflicted on the main contact. In the second class of circuit breaker two series of contacts are also provided, but there is no magnetic blow out arrangement attached to either. In place of this the auxiliary contacts are formed of carbon blocks which are not greatly harmed by the arc, and, in any case, are easily renewable. In closing a circuit breaker the auxiliary contact always comes into action first ; in short, it is the contact that does all the work so far as opening and closing are concerned, the main contact is a shunt to it and is meant only to carry the greater part of the current when the switch is closed.

The design of electric motors has now-a-days become to a great extent standardised, and methods of construction are much the same in all makes. The field frame is usually a circular soft steel casting of which the inwardly projecting poles are sometimes an integral part. In some forms the poles are of a special soft iron, and in others they are built up of a number of soft iron sheets bolted together, the object of this latter construction being to prevent the induction of wasteful eddy currents in the poles. The magnet coils are always wound on formers in a machine, and are slipped on to the poles, being held tightly in position by suitable supports. The armature core is built up of numerous thin soft iron sheets which are clamped together by end plates and keyed to the shaft. The laminations are insulated from one another by a coating of japan or some insulating substance in order to prevent the formation of eddy currents which would otherwise be induced in the core by its rotation in the magnetic field. At intervals, radial air channels are formed in the core to allow of circulation of the air and dissipation of the heat generated in the coils by the current. The armature coils are former wound and embedded in slots punched in the core discs parallel to the shaft, an arrangement which renders the replacement of a faulty coil an easy matter as compared with the old method of hand winding of each separate coil *in situ*. The commutator is built up of hard drawn copper bars insulated from one another and from the shaft by sheets of mica. The segments are held together by end plates shaped to fit into grooves in the ends of the former and clamped up tight. The brush holding mechanism varies greatly, but it is always designed to secure an uniform pressure of the brush on the commutator throughout the whole life of the former, and to allow of easy adjustment, if necessary, when the machine is running. Brushes are now always made of carbon, the results obtained with these being better than those given by the old-fashioned copper brush.

Bearings are usually carried by the end shields bolted to the frame, or at any rate they will constitute a part of the shields unless they spring from the bed-plate, which is not usual. The type of a motor, whether open, semi-enclosed, or totally enclosed depends upon the end shields. The latter class of motor will be larger for a given output than either of the former, as an equal temperature rise cannot be allowed owing to the comparative lack of ventilation. The semi-enclosed motor is the most popular and for ordinary purposes the best of the three.

CHAPTER VI.

THE ELECTRIC MOTOR.

Development of Alternating Current Motors—Classification—The Synchronous Motor—The Asynchronous Motor—Construction and Theory of the Induction Motor—The Single-Phase Induction Motor—Starting and Speed Regulation of Induction Motors—Comparative Efficiency of Direct Current and Polyphase Induction Motors—Frequency and Pressure—The Single-Phase Repulsion Induction Motor—The Single-Phase Series Railway Motor.

THE commercially successful alternating current motor is a product of comparatively recent times. Its development was doubtless retarded by the fact that for years there was but little call for it, the direct current motor having been found reliable and efficient, and capable in every way of filling the somewhat restricted sphere at one time allotted to electric power. With the advent of distribution over larger areas, and the consequent need for alternating currents, the alternating motor began to emerge from the experimental stage; and the comparatively recent introduction of polyphase currents, and their now widespread use both by British and foreign power companies, created the demand which has resulted in the supply.

Primarily, alternating current motors may be divided into two classes: synchronous and asynchronous; each of which may again be divided into single and polyphase, and still further into variants of these. The first class, the synchronous motor, is as old as the alternating current generator itself, since any machine of this type will work as a motor when supplied with current of the same pressure and frequency as that which it would generate at any given speed. This, indeed, is an essential for the parallel running

FIG. 10.—STATOR OF THREE-PHASE MOTOR, SHOWING COILS EMBEDDED IN SLOTS IN THE CORE.

of alternators, that is, the connection of two or more to the same 'bus bars, as only by the motoring action is it possible for independently driven alternating current generators to maintain the exact phase relationship necessary for parallel running. The great disadvantage of the synchronous motor is that it is not self-starting, but must be run up to speed by an auxiliary and then synchronised in the same manner as a generator. Further, being synchronous, its speed can only be altered by a change in the speed of the generator, and its field magnets require a supply of direct current for their excitation. Where synchronous motors are used—and for heavy constant speed work they have certain advantages— they are each provided with an exciter coupled direct to the shaft, which is first used as a motor for running up to speed, and then as a generator for exciting the fields. In such an installation there must of course be a direct current supply of some sort in order to run the exciter as a motor, and this would generally be obtained from a small battery of accumulators installed for that purpose. On the face of things it is obvious that the synchronous motor of this type must possess a very restricted sphere, as its characteristics render it wholly unsuited for every day work. The asynchronous motor, on the other hand, is to a certain extent similar in its behaviour to the direct current, and in some respects—notably in that of simplicity—is its superior. A three-phase asynchronous induction motor (induction for reasons which will become apparent) has two essential parts : the stator, and the rotor. The former, which, as its name implies, is the stationary element, consists of a ring built up of soft iron laminations on which the winding is placed in tunnels or grooves. The three wires from the three-phase circuit are connected to this at electrically equidistant points through a switch, and therefore a three-phase current circulates in the coils. The rotor or rotating element consists of a core of soft iron discs tunnel wound with comparatively heavy conductors, which, in the simplest

form of motor, are all connected together at their ends and
have no connection at all with any circuit external to the
machine.   This type of winding is termed "squirrel cage,"
because in the earliest induction motors it took this form,
consisting of a number of copper bars fixed at both ends into
discs of copper.

Returning to the stator, the effect of the three-phase

FIG. 11.—ROTOR OF THREE-PHASE MOTOR.

current circulating in its windings is to produce a magnetic
field, the poles of which rotate at an uniform speed round
the inner periphery of the core.   This action is similar in
effect to that which would be produced by the mechanical
rotation of a series of magnets.   The cause is not difficult
to follow.   It must be remembered that the three-phase
currents actually lag behind one another in time, and that
one-third of the stator winding is fed by each of these.
The magnetic effect of each winding will, therefore, be
separated by a definite space of time.   Thus, the coils fed
by Phase A will at one moment produce, say, a north pole

at a certain point in the core; a fraction of a second afterwards, the coil connected to Phase B produces a north pole further round the core, and the pole generated by A dies away. Phase C then takes up the work and also produces a north pole, while that of Phase B subsides. The cycle is now complete; and to all intents and purposes the pole originally produced by Phase A has actually travelled along the inner periphery of the core in a manner which may be compared to the mechanical movement of the north pole of a magnet. Granted it is understood that the magnetic field does travel round the core, it follows that in doing so it cuts the rotor conductors, and since these form a closed circuit of low resistance, induces as it passes a considerable current in them. The magnetic effect of the induced current in the rotor core is a pole of opposite sign to that in the stator, and since unlike poles attract, the rotor begins to move. The action of the two-phase induction motor is similar to that of the three-phase, and the mechanical structure is the same in principle.

The foregoing is not offered as a complete explanation of the operation of polyphase motors. There are other actions and interactions which would require consideration if the subject were gone into fully, but these would be out of place in a concise general treatise and in all probability confusing. If the general idea is grasped from the above, its purpose has been fulfilled.

The single-phase induction motor although thoroughly serviceable is not on the whole so satisfactory as the polyphase. It has not the same powerful starting torque, and, moreover, will not carry the same amount of overload. The reason for both these deficiencies lies in the fact that the magnetic field produced in the stator oscillates backward and forward instead of rotating, with the result that the impulses under which the rotor turns are not uniform, as in the case of the polyphase machine. A single-phase motor of this type cannot be started by simple connection of the stator winding to the mains. It is necessary to produce a rotating field in the stator, and this is effected by the addition of an auxiliary winding which is used only at starting, being cut out when the motor is up to speed. The second winding is connected to the supply through an inductive resistance, that is, a resistance which by its coiled shape allows the magnetic field evolved by the current flowing in it to cut through the coils at each alternation and thus generate a back E.M.F. The effect of this back E.M.F. is to cause the current passing through the coil to be displaced backward in phase, and although its frequency will of course be the same as that of the supply from which the coil is fed, the current will be found at any one moment to lag behind that flowing in the main circuit. When such a current is introduced to the auxiliary winding of a single-phase induction motor we have at once a two-phase effect, and therefore a rotating field in the stator. Under these conditions the motor is self-starting, but as the torque is poor it is usual to start on a loose pulley, throwing the load on when the rotor has attained full speed and after the auxiliary winding has been cut out.

In the starting of an induction motor of any description the current taken by the stator winding is momentarily very great, and in all but the smallest sizes it is usual to provide means, similar in effect to the direct current starting switch, for reducing this current to reasonable limits.

There are two principal methods employed for this purpose, both of which are also used for speed regulation. In the first the voltage applied to the stator winding is varied by aid of a transformer. At starting, the transformer is so regulated as to apply a low voltage to the stator, and this is gradually increased as the machine gathers speed, the transformer being finally cut right out of circuit and the whole voltage of the mains applied to the winding. In the second method the ends of the rotor coils are not connected together directly, but are brought out to rings (three in the case of a three-phase motor) fixed on the shaft. Brushes bear upon these rings and connect the rotor windings through wires to a variable resistance external to the machine. At starting, the current induced in the rotor is comparatively small, as the circuits of the coils are completed through resistances which cut it down. The magnetic reaction of the rotor currents on the stator is, therefore, less than it would be if the rotor currents were large, with the consequence that the stator does not take so much current. The effect is identical to that of the alternating current transformer, which, it will be remembered, automatically keeps the current in the primary in proportion to that flowing in the secondary. In the induction motor the stator winding corresponds to the primary of the transformer, and induces current in the rotor which corresponds to the secondary. As resistance is cut out of the rotor circuit the machine increases its speed, and *vice versâ*, hence, speed regulation is easily effected by this means. Compared with the method of varying the speed of a direct current shunt wound motor the arrangement is not economical, and also, it does not give so wide a range;

but the advantages which the polyphase motor possesses **for**
some purposes over the direct current compensate, at least,
for this. The construction is simpler, as there **is no**
commutator, and the motor is not harmed by dirt, **iron**

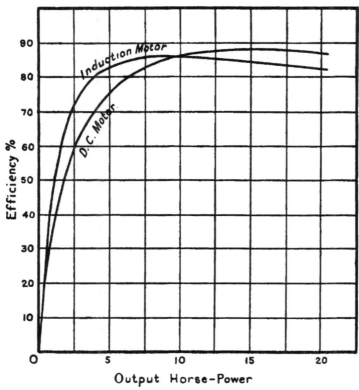

FIG. 12.—CURVES CONTRASTING THE EFFICIENCY AT VARIOUS
LOADS OF 15 H.-P. POLYPHASE INDUCTION AND DIRECT
CURRENT MOTORS.

filings, or other foreign matter. It is an impossibility
under any circumstance for it to race, which is not the case
with any direct current motor ; and repairs to its windings
are easier. Further, the starting and controlling gear is
simpler and less likely to get out of order, and the smaller

sizes, in many cases, require no starting gear at all. In contrasting the characteristics of the direct current and polyphase motor it must be remembered that they each have their own sphere in which the one is preferable to the other. Neither can be said to be the better all round motor; their comparative value depends upon the circumstances under which power is required. Where a wide range of speed is a necessity, and where the motor is not likely to receive very rough treatment or be subjected to the deleterious action of dirt and damp, the direct current machine may be considered preferable. But in situations such as collieries, where widely varying speed is not so essential as it would be in a machine shop, for instance, and where the treatment the motor receives is often of the roughest, the polyphase machine is, for the reasons above stated, deservedly the most popular. In comparing the two efficiency does not enter largely into the question, since in this respect there is little to choose between them. Fig. 12 contrasts the efficiency of a constant speed 15 h.-p. polyphase induction motor with that of a constant speed 15 h.-p. direct current machine at various loads. The two are of the same manufacture, coming from the shops of a well known high class firm. It will be seen that at light loads the induction motor is the more efficient; but at half load this comparative advantage begins to fall off, with the result that the direct current machine is at full load the better motor. It should be noted that these curves are for constant speed; in variable speed work the direct current motor will show up to much better advantage, owing to its more efficient method of speed control.

From 25 to 50 cycles per second are the usual frequencies

employed for the operation of induction motors; at higher periodicities they are not so satisfactory. There is hardly any limit to the voltage, for the reason that the portion of the machine connected to the mains—that is, the stator— is so entirely free from mechanical complications that very little difficulty is met with in the insulation of its con- ductors. Large polyphase motors are often worked direct from the mains at high pressure; an economical arrange- ment since it obviates the losses inherent upon transforma- tion of the pressure. This is another point of superiority over the direct current motor, which, owing to the fact that it possesses a commutator and rotating armature, presents almost insurmountable difficulties in the way of efficient insulation for very high pressures.

Before leaving alternating current motors it is necessary to draw attention to another class of single-phase motor, which of late years has come considerably to the fore. With the ordinary single-phase induction motor the starting torque is poor, and this constitutes one of its chief disadvan- tages. Now it has been found that if what is practically a direct current armature, complete with its commutator, be substituted for the ordinary rotor, and if the brushes bearing on the commutator be short-circuited, the motor will start with a good torque upon connection of the stator windings to the supply. Such a machine is termed a "repulsion motor," because rotation is set up by the repulsive action between the stator field and the field due to the current it induces in the armature. A motor of this class may be provided, at the end of the armature remote from the commutator, with two slip rings connected to symmetrical points in the armature winding. When it has attained full speed a switch is thrown over, which short-circuits the rings and allows the machine to run as an ordinary single-phase induction motor. The particular machine designed upon these lines possesses many of the advantages of the direct current motor. It runs at an approximately constant speed within its range of load, and dispenses with the use of a fast-and-loose pulley arrangement for starting. By the addition of an extra winding it may be made to run in the opposite direction by the simple throwing over of a two-way switch.

There are a large number of types of repulsion motors, and finality has by no means been reached in their design. In addition, there is another totally different class of single-phase motor, which for traction work would appear to have a brilliant future before it. As this motor, so far, has only been adapted in practice to railway operation it will be considered under that section.

CHAPTER VII.

THE APPLICATION OF ELECTRIC POWER.

Statistics of Electricity Supply—Comparative Economy of Electric and Steam or Gas Power—Economic Features of the Electric Motor—Reliability—Causes of Breakdown—Installation of Correct Class and Horse-power of Motor—General Economy of the Electric Drive—Inefficiency of Shafting—Loss in Steam Pipes—Speed Variation of Tools—Workshop Generating Plant—Distribution—Individual and Group Driving—Minor Economies of Electrical Operation.

THAT electric power possesses, for most purposes, advantages over gas and steam power is evidenced by the manner in which the motor has of late years ousted its rivals from positions which were at one time thought to be impregnable. Wherever there is a cheap and suitable electric supply, gas and steam plants are daily diminishing in numbers ; and the motor connections of every authority giving such a supply are increasing in a manner which speaks volumes for the all-round efficiency of electric power.

That much remains yet to be done in this country, particularly in London, is evidenced by the following Table (page 67), prepared by the Administrative County of London and District Electric Power Co. at the time their Bill was before Parliament in 1905.

It compares the principal German and American cities, and some of the chief British power areas, with results most unfavourable to the latter, excepting in so far as Tyneside is concerned.

In this enterprising district, despite the cheapness of coal and gas, the advantages of electric power have been more fully appreciated than in any other British area. For London there is some excuse. Quoting again from the Administrative Power Company, in London and the surrounding districts, there are no less than 61—on the average—comparatively small generating stations, supplying on 27 different systems, and at 24 different pressures. The majority are in the hands of municipalities, and financially

COMPARISON OF ELECTRICITY SUPPLY IN LONDON WITH VARIOUS
CITIES AT HOME AND ABROAD.

Name of City.	Population.	H.P. connected per 1,000 head of population.	
		All Purposes.	Power only.
Boston	600,929	164·6	41·3
Tyneside	222,241	122·8	30·0
Frankfurt	306,000	80·1	28·7
Berlin	2,285,000	82·1	22·0
New York	3,732,903	52·6	13·9
Hamburg	700,000	76·7	12·7
Glasgow	786,897	43·9	8·26
London	6,565,390	48·0	5·4
London (Industrial Area)...	3,812,283	25·8	4·8

are not great successes. The average capacity of the works is under 3,000 k.w. ; and the average size of the generating sets 338 k.w. Such conditions are obviously unfavourable in the extreme to the cheap generation of power ; and until this is undertaken by some large central body, equipped with plant comparable, for instance, to that in use on the Tyneside, and under equally efficient management, London must remain behind.

F 2

The comparative backwardness of Glasgow and other prominent cities where supply conditions are on the whole more favourable than in London is due probably to a natural reluctance to scrap existing satisfactory plant. The supply of electric power at rates which enable it to compete with gas and steam is a comparatively new thing, and there is yet much prejudice and ignorance to overcome. Also, although a manufacturer may be convinced of the fact that electric power will prove more economical to him, its installation in many cases will call for a considerable expenditure, which very possibly he is unable or unwilling to incur. There is no excuse nowadays for a new factory to be run by anything but electric power provided the supply is cheap, or failing this, provided current can be generated economically by a plant installed for the purpose. And it is safe to say that under such conditions it is very unusual for anything but electricity to be employed. On the other hand, the ousting of gas and steam plant, with its wasteful lines of shafting, is often a difficult matter to bring about owing to the extra capital outlay involved; and until renewal of the plant becomes necessary, or until the stress of competition renders more economical methods imperative, many factories and works must remain behind the times in their equipment.

Broadly speaking, the economy of electric. power for industrial purposes lies in two directions: firstly, in the motor itself, and secondly, in the system of working and arrangement of plant, which the motor makes possible. It has been pointed out that the motor is a highly efficient machine, particularly when proportioned to its load in a manner that allows it to be run continuously at or near its

FIG. 13.—MOTOR FIXED TO WALL AND DRIVING WOOD-WORKING
MACHINERY.

full output. Beyond this it possesses many good features which are either totally or partially absent in gas or steam plant. It requires no attendance when running, which cannot be said of either of the former. It takes up surprisingly little room; and as attendance is not necessary it may be fixed in any odd and otherwise useless corner. The repairs bill is low, for the reason that there is little to go wrong; and accidents are prevented by trustworthy automatic switching arrangements, such as were described in the preceding chapter. The motor being purely a rotary engine its speed is perfectly uniform, a matter of great importance in some industries; and the efficient and simple manner in which the speed can be varied over a wide range is a point of considerable value. It makes practically no noise, requires no foundations to speak of, and is entirely free from vibration; all of which are characteristics more or less conspicuous by their absence in reciprocating engines. Similarly, the ease and celerity with which it can be started and stopped is unapproached by any other form of machine. This is practically the case for the motor. There is little that can be urged against it. The running cost is sometimes high compared with gas, but this is due, not to the motor itself, but to the high price of current which still obtains in many districts. The remedy is being applied every day; and before long we shall undoubtedly see power universally supplied, even in small quantities, at a penny per unit. In some parts of the country, notably in the north, the price is well below this moderate figure; and when such is the case the motor can compete easily with its rivals. There are many indirect economies attached to the use of electric power, and when these are fully taken into consideration the balance will often be found in its favour, although at first sight the reverse might appear to be the case. In some quarters the motor has been stigmatised with unreliability, and there are instances where it has been superseded by steam and gas plant for this reason. It is impossible to particularise, but generally speaking, when a motor proves unreliable it is from one of three causes. It may belong to that cheap and shoddy class of machine, turned out in thousands in America and Germany to meet the demand which unfortunately exists for such. If this be the case there is nothing further to say; breakdowns and inefficiency can only be expected. On the other hand, the motor may be of a first-class grade and yet prove a constant source of trouble and loss. If so, probably one, or perhaps both, of the two remaining causes are present. The machine may be systematically ill-used through ignorance or carelessness, or it may be totally unsuited by its size or type for the class of work it is used on. A good motor is a thoroughly mechanical machine and essentially robust in its nature; but its powers to withstand a perpetual state of dirt, for instance, are limited. The brushes of a direct current motor, when once set, require little attention beyond very occasional renewal; but judging from experience even this is sometimes denied them, and for the want of some slight adjustment, comparable perhaps to the tightening of a nut on an engine, brushes are left to spark and reduce themselves and the commutator to a state in which they cannot perform their functions properly. As an example of the other extreme, many an armature has been returned to the makers with its commutator worn right through owing to the well-meant but mistaken attentions of a person whose knowledge of electrical matters went not beyond appreciation of the value of a good contact. A reciprocating engine will always have its attendant, who will be more or less versed in its treatment; a motor, because it requires so little attention, sometimes gets none at all, or is entrusted to the hands of those utterly unfitted to take charge of machinery in any form.

The third main cause of trouble with motors is one which need not be looked for where a good firm have been employed to carry out the installation. It can only arise out of the inexperience of amateurs dealing with a subject which requires considerable specialised knowledge. Wherever motors are installed the work should be carried out by a firm duly qualified, or under the direction of a competent and experienced engineer. If this course be taken, breakdown and general unreliability need not be feared. There is often great difficulty in ascertaining the horse-power required to drive machinery ; and it has frequently happened where, for instance, some heavy tool has been disconnected from shafting and adapted to a separate motor drive, that the motor provided has proved too small. After working continuously with a heavy overload it breaks down ; and it is at once assumed that the cause lies in the motor itself and not in the way it has been used. Such an occurrence is possible where a manufacturer (knowing nothing of the subject) buys a motor and instals it himself without having made proper inquiry into the matter. But no first-class engineering firm will risk its reputation in this way. In a like manner, the wrong class of motor may be provided, shunt wound instead of series, perhaps ; or, the characteristics may not be suitable; for motors, although similar in class, are often constructed in such a manner as to give rather different results. For the uninitiated there are many pitfalls in the purchase of a motor unless under skilled advice; if this were always taken the stigma of unreliability would never be attached to the good class of motor.

Although in itself often a source of economy, it is the conditions rendered possible by the electric motor which constitute its chief claim for consideration. To understand the advantage of these fully it is necessary first to review the conditions under which a steam- or gas-driven factory is run. Assume that there is one main engine of sufficient horse-power to drive the whole factory. This will distribute power throughout the various departments by the agency of belts and shafting. Where machines require varying speed it will be obtained by pulleys of different diameters or by some form of mechanical gearing. Starting and stopping of the machines will be effected by shifting the belt off or on to loose pulleys or by throwing in or out the gear. It will be obvious that such conditions are by no means ideal. To start with, the engine must be kept running if only one machine or tool be required; and although at full load it may be highly efficient, its light load working will be comparatively uneconomical. The machine in use may be situated some distance from the engine, in which case a great deal of shafting and belting will have to be kept in motion for no useful purpose but to transmit the power. Such light load conditions of working are by no means uncommon, especially in engineering shops where overtime is frequently worked to complete some small job. We have therefore one set of ordinary conditions under which factory operation must be highly uneconomical on this system. But now consider things under their best aspect; that is, when the engine is working economically at full load and the whole workshop is in operation. Still there is great inefficiency, for the reason that the shafting and belts absorb power. The old-fashioned engineer who clings tenaciously to the methods of his boyhood will often maintain that this is, if not inappreciable, at the most of little account; but it has been proved over and over again, by the most simple expedient of indicating the engine with nothing but the shafting running, that this loss is not only appreciable but in some cases is large enough to absorb the greater part of its output. Necessarily, conditions materially affect results; but in most shops of any size it is within the mark to anticipate a 10 per cent. loss of power in shafting, and cases have been known where the waste has amounted even to 70 per cent. To give a few examples : In a cotton mill driven by a steam engine working most of its time about full load, the engine and shafting together accounted for a 25 per cent. loss of power—by no means a bad result for mechanical distribution, but one which was attained under favourable conditions. An engineering shop driven by a steam engine on which the load was variable showed a 48 per cent. loss; and another similar shop run by a gas engine, which most of the time worked at half load, wasted 50 per cent. of the indicated horse-power of the engine. Many worse examples than these can be found ; for instance, a large printing works was proved to waste 56 per cent. of the power delivered to the main driving belt in shafting and gearing alone. It is, however, hardly necessary to cite further instances to prove the inefficiency of shafting and

mechanical distribution generally; the point has been con-
ceded by all in a position to know. Efficiency may of
course be improved in one direction by the splitting up of
shafting and the installation of small engines at suitable
points. But if these are steam engines a further loss appears,
namely, that which occurs in long lengths of steam pipes;
and in addition, the efficiency of the engine itself will be
low compared with that of the larger set. With gas engines,
of course, there is an improvement, as the loss by condensa-
tion in steam pipe is eliminated; but other considerations
are present which, as shown, still leaves the balance of
all-round economy with the motor. In engine-driven works
the range of speed obtainable on the machines is coarse and
limited unless a large number of pulleys varying widely in
diameter are installed. Under these conditions, speed variation is a slow process, as belts have to be shifted from one set
of pulleys to another; and unless the belt slips considerably,
starting is abrupt to an extent that is not always desirable.

Comparing the working of an electrically-driven factory
on any one of the above points, which are by no means an
exhaustive list, we find an improvement in each.

Where the price per unit is low enough it is generally
preferable to take supply from the mains, if only for the
reason that there is less risk of failure. It is, however,
very usual in a large works to instal a special generating
plant, which, particularly in the case of engineering shops,
may usually be so designed and worked as to result in very
low cost of generation. Such plant will be similar to that
of a small supply station. It should include a battery for
use on light overtime work, thus enabling the generators to
be shut down; or as an alternative, there should be a

throw-over on to the outside supply mains, an arrangement desirable from the point of view of safety as well as the above.   From the switchboard the feeders will radiate to all parts of the works, taking the place of main shafting in the mechanically-driven shop and wasting at full load only about 3 per cent. of the power transmitted.   At light load the waste will be very much less owing to the smaller drop of pressure in the cables.   Obviously, by the electrical system power may be delivered at any part of the works irrespective of distance, in a highly economical manner. Further, the cables require no attention and take up no useful room, which reduces maintenance costs and makes for general convenience.   Each large machine or tool will be fitted with its own motor, and small apparatus will be grouped together on countershafting.   Opinions are divided on this point, and the grouping of machines must remain very much a matter of opinion or expediency ; but although the individual motor drive possesses many obvious advantages, particularly that of flexibility, it must be remembered that a small motor is not so efficient as a large one, and also, that to instal a motor to each machine is an extremely costly method.   It has been pointed out that efficient and delicate speed variation is readily obtainable with the electric motor, and here there is a very obvious advantage over mechanical methods.   In the same way, starting may be as gradual as is wished.   In small shops where only one main prime mover can be employed the losses in shafting, etc., are of course the same whatever form of power is used.   For this class of work the advantage of electric power lies in the greater efficiency of the motor, especially at light loads, and in the fact that there are no stand-by losses as with the steam-engine. Minor economies may also be effected in the general operation of the plant; for instance, being so readily started, the motor may be shut down for a few minutes only if necessary, whereas a gas engine would be left running on a loose pulley because of the difficulty of starting again. Further, attendance and repairs are both lower, the space taken up is less, and fire insurance will be quoted for at a lower rate than where steam or gas engines are installed. In a large workshop with group driving the electric motor results in better sanitary conditions for the hands, a comparatively clear overhead space owing to the absence of most of the usual shafting and belts, for the same reason better light and less noise, and greatly reduced risk in operation of the plant. Buildings of a less substantial and consequently cheaper nature are required; and the tools or machinery may be arranged in the most convenient manner for cheap production instead of being restricted to a limited number of positions. Isolated buildings are supplied with power in a much more economical manner than is possible with mechanical methods, and extensions can be effected in any direction without stopping the works for a moment, and in the simplest manner.

The use of electricity has resulted in the evolution of a new class of tool which by its portability and independence of shafting can be brought to the work instead of the work to the tool. The latter course, it should be noted, is often an expensive operation where heavy weights are dealt with. Further mention will be made of such apparatus later; enough has been said to show that the electric drive in some suitable form provides a certain means for cutting down expenses and, concurrently, increasing output.

Unfavourable Conditions Existing—Improvements effected by Electric Power—Home Office Rules—Comparative Value of Induction and Direct Current Motors—Pressure and Frequency—Distribution of Supply — Cables — Substations — Precautions against Fire — Machinery requiring Power—Winding—Haulage—Locomotives —Drilling—Air Compressors—Coal Cutting—Pumping—Ventilating.

THE use of electric power in collieries and mining work generally has of late years received a considerable impetus, and it now stands by itself in the position of a highly specialised branch of the electrical industry. The conditions under which work is carried on in a colliery are by no means favourable for electric plant. Dirt, damp, and rough usage have all to be anticipated, and for this reason the plant must not only be well constructed, but it must be designed to a great extent upon lines calculated to render it specially suitable for the extraordinary conditions which obtain. As elsewhere, the introduction of electric power has in many cases resulted in enhanced convenience and lower costs. Wasteful rope drives and long pneumatic pipe lines have been abolished, hand labour has been superseded, greater speed and safety in the operation of machinery has been attained, and, generally, conditions have been in every way improved. An idea of the difficulties and risks encountered in colliery work will perhaps be best attained by a perusal

FIG. 14.—GAS-TIGHT MOTOR STARTER AND REGULATOR FOR USE IN MINES.

of the Home Office rules for the installation and use of electricity in coal mines. All work has to be carried out in accordance with these rules, which provide that the following conditions, among others, should be fulfilled. Where the generating station is not within 400 yards of the pit mouth a switch box or house must be placed near the latter, containing main switches controlling the whole of the supply to the mine. By this means current can be entirely cut off on the spot in emergency. All cables must be protected, highly insulated, and substantially fixed and supported ; and where there is danger of fire damp, joints must be made in suitable boxes. Much importance is attached to switch gear, of which the live parts must be protected by covers of an incombustible material unless the gear is placed in special chambers. In fiery portions of the mine, all switches and fuses must be enclosed in gas-tight boxes, or must break circuit under oil. Fuses and circuit breakers must be adapted to open when the current exceeds the working current by 200 per cent. The regulations where motors are concerned are also stringent and indicate the dangers which have to be provided against. All motors and their starting gear must be protected by switches fixed in close proximity that will entirely cut the current off them ; and every motor of 10 h.-p. or over must, if situated underground, be provided with an ammeter to check the load on it and so prevent breakdowns through overloading. The leads passing into the frame of the motor must, where necessary, be bushed in such a manner as to be gas-tight ; and in dangerous positions the motors themselves, their starting switches, terminals and connections, must be completely enclosed in flame-tight enclosures of sufficient strength to resist any explosion of fire damp which might occur within them. Where the presence of gas is likely, a safety lamp or other device for detecting fire damp must be provided for use with each machine when working; and in the event of this indicating danger the motor must at once be stopped and the current cut off· Motors are not allowed to be run continuously for a period of longer than one week unless an inspection of the machine be made at the close of the run to see if it is in a safe condition; and cleaning must wherever possible be thoroughly carried out once a week. Coal cutters and drilling machines must not be left to work unattended, and when stopped, current must be cut off the flexible leads connecting them with the supply in case these should be injured and give rise to fire. Any electrical machine emitting a spark or arc must be stopped at once and the defect remedied before it is started again. Overhead trolley wires are not permissible for the working of mine locomotives where there is risk of fire damp, owing to the possibility of the gases becoming ignited by a spark between the wire and the trolley.

It will be seen that everywhere the first precaution is that against fire and explosion. In a coal mine whatever can be done to minimise the risk to which employees are subjected in this respect must be done; and it must take precedence over all questions of convenience or economy. Safety is the first consideration, cost the second; and for this reason the polyphase motor—which, having no commutator, does not spark—is generally considered preferable. It has also other advantages, chief among which are its more mechanical structure and comparatively simple switch

E.P. G

gear. Compared with the direct current motor, it is not so economical or capable as a variable speed machine, which is of course in many cases a disadvantage; but the former is always a source of potential danger owing to sparking at the brushes, and no means have yet been found for making a motor thoroughly gas-tight. A polyphase induction motor is more easily and quickly reversed than a direct current machine, and if heavily overloaded will pull up without danger of burning out, both of which are valuable features in colliery work. It will be seen, therefore, that the principal advantage of the direct current motor lies in its efficient and wide range of speed control; further, it is undoubtedly preferable for general work and for the operation of mine locomotives. For the above reasons both systems are sometimes installed in the same mine, each being used for the driving of that class of machinery for which it is most suited.

In many collieries the workings or positions where power is required are a considerable distance from the generating station. When this is the case the polyphase alternating current system possesses the great advantage of easy and efficient transformation, and therefore renders possible the use of high tension transmission and low tension distribution through the agency of sub-stations. A three-phase alternating current of 2,000 or 3,000 volts pressure and a periodicity of 25 or 50 cycles is frequently employed. Power may be taken in bulk from some adjacent supply company, but many collieries, particularly the larger ones, possess their own generating station, which will be equipped upon the usual lines. Wherever possible the current is conveyed to the pit mouth by overhead wires for the reasons of economy which have been gone into, but it is conducted down the shaft by insulated and armoured cables in duplicate, designed and fixed in a manner suitable to meet the somewhat unusual conditions which obtain. In the case of a deep pit the weight of the cable is very considerable, and it has therefore to be well anchored at the top and securely clamped to the wall in a large number of places, generally every fifteen feet of its length. Lead covering, although it effectually excludes damp, has been proved not to be desirable; but armouring of some sort is essential in order to preserve the cable from mechanical injury. Both galvanised steel wire and helical steel tape wrappings are used, but the latter is not so satisfactory as the former, as its power to resist tension is much less. For three-phase high pressure transmission a three-core stranded cable with waterproof covering of dialite or bitumen and a heavy braiding over the armouring is very successful. In addition to the armouring the cable is sometimes protected by tight-fitting grooved casing or iron pipes. The sub-stations are situated at the bottom of the shaft or other suitable positions in the workings. Where it is necessary to run high tension cables along the main roads they are either buried underground in pipes or suspended from the roof. In the latter case, at places where there is danger of a fall they are left slack, by which the probability of injury from this cause is minimised.

The equipment of the sub-stations, consisting of transformers and switch gear, will present little out of the ordinary beyond the greatest precautions against fire. Here the high tension current will be converted to a pressure of approximately 500 volts for power and 100 volts for lighting; and distribution to main parts of the mine will be carried out by protected cables fixed in a manner suitable for the particular conditions. Care has to be exercised in this respect ; the chief dangers to be guarded against being an overheated and damp atmosphere, rough usage, and the acid state of the water with which the cables will frequently come in contact.

In a colliery, power is required, roughly speaking, for two classes of machinery: that above ground and that below the surface. The winding gear situated at the top of the pit is used for bringing the coal to the surface and for raising and lowering the men. Either direct-current or polyphase motors may be employed for this purpose, and the power necessary may vary from 50 to 2,000 h.-p., according to circumstances. Where the amount called for approaches the latter figure it is usual to set aside one generator in the power house to deal specially with it. The work performed by winding engines is intermittent with frequent short stops, and if the large amount of power required were provided by the main generating plant the effect on the pressure regulation of the system would be very marked, so much so as seriously to hamper the operation of the other machinery. Various methods have been devised to overcome this difficulty and allow the winding gear to be driven from the main plant. The point aimed at is to provide arrangements which have a damping or fly-wheel effect on the varying load, and several of these systems are in use with satisfactory results. Either direct current or polyphase motors may be used for winding, but slow speed is in most cases an essential. High speed motors may of course be employed in conjunction with speed-reducing gear ; but although by this means the capital cost is lessened, the gearing introduces inefficiency and another point where breakdown may occur.

Rope haulage is used in a colliery to convey the coal from the working face to the bottom of the pit.   In this system

FIG. 15.—COLLIERY HAULAGE GEAR EQUIPPED WITH THREE-PHASE MOTOR
AND CONTROLLER.

an endless rope is usually employed in connection with tubs or small wagons running on rails, the motive power being supplied by a prime mover of some sort over the pulleys of which the rope runs.   That much power is lost by such a method goes without saying, and in the case of mines where the workings extend a considerable distance the inefficiency is such as entirely to preclude the driving of the rope from one main engine.   For reasons of safety it is

generally impossible to instal a series of small engines
working the ropes in sections, and it becomes necessary to
fall back on the electric motor, which is eminently suited for
the circumstances.   Pneumatic power may of course be
employed, but it is, comparatively speaking, inefficient, and
troublesome to maintain owing to the leaks which per-
petually occur in the pipe line.   The polyphase motor is
well suited for this class of work, as it may be placed in
fiery portions of the mine with no risk of explosion through
sparking.   The direct-current motor would only be used
where no such danger is to be anticipated.   The power
required may vary from 15 to 500 h.-p.; and it is necessary
that the motors should be able to exert a good starting torque
and to withstand rough usage.

For hauling work, electric locomotives are also used, and
they possess the advantages over rope hauls of greater
flexibility and efficiency.   On the other hand, they introduce
certain difficulties which under some circumstances prove
insurmountable.   The bare overhead trolley wire is a
source of danger from shock; and for this reason it must
be placed out of reach.   A risk also arises from sparking
at the trolley; and as this difficulty can never be completely
got over, the locomotive is unsuitable for fiery mines.
Further, the maximum grade upon which it can be used
is approximately 1 in 20; whereas the rope haul is
hampered by no such restriction.   These locomotives are
built with a narrow gauge, and are kept as low as possible
on account of the small head room.   They are usually
equipped with two series-wound direct-current motors and
vary in size from 15 to 100 h.-p.   The controlling gear is
similar to that used on tramcars; and the whole of the

apparatus is covered in with a heavy iron shell, capable to a great extent of excluding dust and of withstanding the rough usage it is liable to meet with. The colliery loco-motive is more in evidence in America and Germany than in this country, owing chiefly to the somewhat different conditions which obtain.

Drilling is the operation to which it has been found most

FIG. 16.—ELECTRICAL MINE LOCOMOTIVE.

difficult to apply electric power, the reason being that electricity does not lend itself readily to the production of the reciprocating or percussive effect which for this purpose is preferable to a rotary one. The difficulty is overcome by the installation at suitable positions in the workings of small motor-driven air compressors which furnish air under pressure for the operation of pneumatic drills. This system is also preferable from the point of view of safety, as it introduces no risk of explosion at the working face through an arc. Both polyphase and direct-current motors are suitable for the driving of compressors, the motor being compound wound in the latter case.

Coal cutting by electric power is one of the most recent developments of colliery work. Such machines absorb from 20 to 50 h.-p. and differ considerably in form according to the purpose for which they are used. Dust-proof construction of the motor is here more than ever necessary, and it is also essential that the casing should be as far as possible gas-tight. Owing to the risk of ignition through sparking, the polyphase motor is preferable for this class of work, but the direct-current machine is the more easily adapted to it. The load on the motor is very variable, and therefore apt to cause sparking in the direct-current machine.

Pumping is an operation for which the electric motor of either class is particularly suitable. The load is steady and allows the motor to be run continually at its full capacity and therefore in the most efficient manner. The draining of a mine by the aid of steam pumps is seldom a satisfactory method, as for obvious reasons the steam plant must be to a great extent centralised and placed always where there is no danger of explosion. On the other hand, the motor-driven pump may be installed just where it is wanted and in the most favourable position for it to do its work, efficiently. Furthermore, it is compact and in small sizes readily portable—both great conveniences in some classes of work. The power required may vary from 5 to 500 h.-p.; and where direct-current motors are used the compound-wound machine is the most suitable.

Considerable progress has been made in efficiency and

ease of ventilation since the introduction of electric power. The crude method of placing a fire at the bottom of the shaft to induce a flow of air was improved upon by the employment of steam-driven fans situated near the mouth. But this arrangement, while satisfactory in effect so far as the main portions of the mine are concerned, is obviously unable to deal properly with what may be termed local ventilation, for which the electrically-driven fan is the only possible solution. In the working of a colliery main ventilating plant the economic advantages are entirely upon the side of motor-driven fans. They have no stand-by losses such as would be incurred with a separate steam-driven plant, and they require very little attendance even when the installation runs into hundreds of horse-power. Their speed and output of air is variable economically over a wide range ; and should the maximum power be required suddenly in emergency it may be attained in a few seconds without any of the preparation which in many cases would be necessary with steam plant. Ventilation is of such vital importance in a mine that it is incumbent upon owners to instal the best possible system. Experience has proved that this consists of a combination of electrically-driven main fans of large size, and smaller ones (preferably of high speed) installed in suitable parts of the workings.

CHAPTER IX.

ELECTRIC POWER IN ENGINEERING WORKSHOPS.

Individual and Group Driving—Direct- and Alternating- Current
Motors—Fixing of Motors—Horse-power and Type of Motor
required for various Tools—Multi-voltage Systems of Speed Con-
trol—Electric Travelling Cranes—Comparative Value of One and
Three Motor Cranes—The Induction Motor for Crane Work—
Jib Cranes—Derrick Cranes—Lifting Magnets—Magnetic Sepa-
rators—Portable Tools—The Hand Drill—The Electric Pulley
Block.

THE application of electric power to an engineering workshop is a wide subject and one full of interesting problems.
To obtain satisfactory working it is necessary to consider
the requirements of each individual machine or tool, and to
supply the right class of motor and instal it in the right
way. The question of individual and group driving has
been touched upon. In the class of work under consideration
the maximum of economy is obtained by a judicious combination of the two; machines requiring over 5 h.-p. being
driven independently by their own motors, and smaller tools
being grouped together and run from one motor by the aid
of countershafting and belts. The arrangement of these
latter must be very carefully thought out, only such tools
as are liable to be in operation at the same time being
grouped together. The 5 h.-p. rule does not, of course,
apply to portable tools. Many of these will require a lower
power; but the convenience and saving effected by not
having to move heavy work from place to place in the

FIG. 1. LARGE LATHE DRIVEN BY THREE-PHASE MOTOR.

shop more than compensates for the extra outlay incurred and the comparative inefficiency of the smaller motor. Both direct-current and induction motors are used for machine shop work, the former being generally preferred owing to their more economical speed control. In deciding between the two, in the absence of any important considera- tion it is a matter of personal preference, either for wide range and economical speed regulation, or for a motor which will work reliably under the most disadvantageous conditions of dirt and rough usage. In some large works both systems will be found; direct-current motors being applied to tools requiring variable speed, and induction motors to machines operating on a rougher class of work.

The motors used with a group drive may be fixed in any suitable position, either on the floor, wall, ceiling, or on supports specially provided for the purpose. In each of these cases it must be seen that the right type of lubricating arrangement is supplied, as obviously a motor designed for use in an upright position and fitted with loose ring and oil-well lubricating gear would not work in an inverted or right angle position.

Machines designed for individual drive often have the motor fixed rigidly to them, the power being transmitted by a train of gear wheels, or by short belts, or chains. The starting and speed-regulating switches are also commonly fixed to the machine itself, which is thus controlled in the handiest and safest manner possible.

The tools and machines employed in an engineering workshop or shipbuilding yard vary considerably in size and nature. It will be instructive to take a few examples, indicating the class and horse-power of motor which is usually employed in each case. The ordinary screw-cutting lathe requires from $\frac{1}{4}$ to 2 h.-p.; the turret lathe from 1 to 5 h.-p.; and the face-plate lathe, which is used for

FIG. 18.—PUNCHING AND SHEARING MACHINE FITTED WITH 25 H.-P.
DIRECT CURRENT MOTOR AND STARTING SWITCH.

the heaviest work, from 2 to 10 h.-p.   These are average
sizes only; in many instances the horse-power required will
be considerably higher.   In each case speed variation is
necessary, and where the lathe is driven individually the
direct-current motor is therefore preferable.   On the other
hand, where the machines are operated in groups and variation of motor speed is not so essential, the induction motor
is equally suitable, change of speed in this case being
effected by mechanical gearing.   Drilling machines require
from $\frac{1}{2}$ to 2 h.-p., and are generally driven in groups.
Either shunt-wound direct-current or induction motors
are suitable for the work, and in this category may also
be included shapers, slotters, and milling machines of
ordinary size.   Punching and shearing machines, used
for punching holes, shearing plates, bars, etc., consume
a good deal of power, and subject the motor to considerable
strains.   As it is capable of exerting a very large torque,
and is mechanically strong, the induction motor is perhaps
the better for this purpose, variable speed not being of
importance.   From 2 to 25 h.-p. is usually required for
this class of tool.   Planing machines absorb a considerable
amount of power at the reverse, and may call for as much as
40 h.-p.: they are consequently best driven independently.
There is little to choose between the direct-current and
induction motor here; but if the former be used it should
be compound-wound in order to cope successfully with
the occasional heavy demand made upon it.   Plate rolls,
used for bending plates into shape, are best operated by
series wound direct current motors, fitted with a brake to
prevent them racing when the load is removed.   They
frequently require a steady maximum of 20 h.-p., which,

momentarily, may be increased 100 per cent. Independent driving is, of course, necessary, and the series wound direct current motor is preferable because it can best meet the occasional large increase in the demand for power. Induction motors are, however, used successfully on this work, which is obviously of a very trying nature, calling for sound mechanical structure in the motor and a powerful torque.

For some classes of work a wide and efficient range of speed control is so important that special direct current systems have been devised which attain the desired end by the application of different voltages to the motor terminals. Such methods are not popular in this country, but in America they have been largely adopted. Early practice in this direction tended towards the use of three and sometimes four different voltages; but the advent of better motor design has rendered such a number unnecessary, and two is practically the standard. Under these conditions the ordinary three-wire system suits the case very well. To all the tools requiring variable speed the three wires are taken, the working pressure being generally 120 volts between the neutral and either of the outers, and 240 volts between the two latter. The slowest speed is obtained by connecting the motor to either of the 120-volt circuits with the regulating resistance in its shunt completely cut out. To increase the speed this resistance is put in circuit, and by this means a certain maximum is attained. For still higher speed the next step is to connect the motor to the two outer wires—which are at a difference of potential of 240 volts—the regulating resistance being again cut completely out. Further increments, up to the absolute maximum, are secured in the same manner as before, namely, by inserting more of the variable resistance in the shunt winding. A speed ratio of 5 to 1 is the usual limit of machine shop requirements, and this is easily and efficiently reached by the above system.

The most interesting, and at the same time most valuable,

FIG. 19.—THREE-PHASE MOTOR EQUIPMENT OF A HEAVY TRAVELLING CRANE.

machine in an engineering shop is the travelling crane. In no other branch of work is the electric motor more successful ; and its performance in this sphere has done much to introduce the complete electric drive to workshops in order that its admitted advantages for crane operation may be secured. It was the first piece of factory machinery for which electric power was employed, coming as a welcome relief from hand or steam-driven cranes by reason of its superior speed and flexibility of operation. There are two systems of applying power to travelling cranes, the first of which necessitates one, and the second three motors. Confining the subject for the time being to the limits of direct current practice: in the former method a shunt or compound wound motor is installed on the crane, and so connected by gearing as to be capable of performing either one of the three motions required, namely, hoisting, longitudinal travelling, and cross travelling. The motor is kept continually running in one direction, approximately at constant speed, and its electrical equipment is therefore simple. The crane is worked by manipulating the gears, which are necessarily complicated, particularly when they are arranged to give variable speed. The advantages of this system are cheapness and ease of adaption to existing cranes; points which must be taken into consideration where the work is light or irregular. The disadvantages are, liability to breakdown in the gearing, comparative slowness of operation, and lack of flexibility. The three-motor system is undoubtedly the better where the work work warrants the cost of its installation. The electrical complication is greater, since speed variation and reversal of motion are effected by the motor itself and not by gearing. On the other hand, this latter is reduced to a minimum, and its attendant evils are therefore absent. In a crane of this type there is a separate motor for each of the motions, and as all three call for considerable starting torque and a speed of operation which varies inversely with the load, the series wound motor is always used. This alone constitutes a point of preference over the single-motor crane, which being worked by a shunt wound motor, has not the same powerful initial effort and carries out its operations approximately at the same speed irrespective of the load. A light weight should be lifted and carried speedily, and empty tackle should be lowered as quickly as possible ; a heavy weight should be dealt with more slowly. The series wound motor fulfils these condit.ons perfectly ; the shunt wound motor, being a constant speed machine, cannot, ánd gearing has, therefore, to be resorted to with consequent loss of time and inefficiency. As the series motor will race at light or no load and possibly wreck itself by excessive speed, the hoisting motor is always provided with a brake which automatically comes into action when the empty tackle is being raised or lowered. Means are provided for putting the automatic arrangement out of action if desired, so that when a heavy weight is being lowered the brake assumes its natural position and exerts its full effort the whole time. A fourth motor is occasionally installed for dealing specially with light weights at high speeds, but this will only be found in special cases. A three-motor crane is operated by the aid of special controllers, which, although often complicated in their connection. are outwardly simple and fitted for working by unskilled hands. The handles are generally locked in such a manner as to preclude the possibility of a mistake. A travelling crane picks up its current from bare conductors fixed parallel to its whole run. The longitudinal travel motor is stationary, and is fed direct from this ; but the crab and hoisting motors collect current from other bare lines strung from end to end of the crane itself and connected to the main collecting gear through the controllers.

FIG. 20.—JIB CRANE FITTED WITH DIRECT CURRENT MOTORS.

The polyphase induction motor was at one time assumed
not to be very suitable for crane work; but this opinion has
now been reversed, and the alternating motor of this class
is almost in as much evidence as its rival.   Its previously
mentioned characteristics of sturdiness and absence of
commutator troubles stand it in good stead, and, indeed, give
it a certain advantage over the direct current machine.
The controlling gear is far simpler, and in short, there is in
every part of the equipment less to go wrong.   It shares
with the direct current motor the power of returning energy
to the mains when lowering a load, by reversing its operation and acting as a generator.   But, unlike the series
wound direct current machine, its speed can never become
excessive.   So far as a comparison of their practical working goes, there remains little to be said.   Taking all points
into consideration, both motors are eminently suited to the
work, and both can show improvement in efficiency and
convenience over any form of mechanical power.

Beyond the ordinary workshop travelling crane, there are
many other types, in all of which electric power has been
successfully employed.   The small jib crane, stationary or
otherwise, is usually fitted with one motor; the operations
of hoisting, slewing—and if movable—of travelling, being
effected by the aid of clutches and gearing.   Heavy derrick
cranes, such as are employed in docks, form the opposite
extreme, and are equipped with motors for each action
combined with comprehensive controlling apparatus.   A
50-ton derrick crane, for instance, will usually be fitted with
five motors, either series wound direct current, or of the
polyphase induction type.   In a crane of this class and size
supplied by Messrs Stothert and Pitt to the London and

South Western Railway Co., there are two 50 h.-p. hoisting motors arranged to work either in series or parallel on 480 volts direct current. The system of series-parallel control will be found explained in detail in the traction section, to

FIG. 21.—ELECTRICALLY-OPERATED 50-TON DERRICK CRANE.

which it primarily belongs. For the present it will suffice to state that this system of control avoids rheostatic losses to a great extent and is, therefore, more efficient than one which employs a plain resistance for starting and speed variation. There is one derricking motor of 80 h.-p., and travelling and slewing are provided for by 50 h.-p. and 25 h.-p. motors respectively.   The speed of the motors is reduced by spur wheels, and the hoisting and derricking gears are equipped with magnetic brakes.   The controlling apparatus is placed in a cabin erected in a commanding position on the front of the crane, the entire working of which is carried out by one man.   The weight of the crane, including ballast and with the load on, is approximately 375 tons.

A piece of apparatus which has been found an extremely handy auxiliary to the crane is the lifting magnet.   It is largely used in the foundry and shipyard for handling iron bars, billets, plates, etc., which are not always easily slung. The saving in time is very great, since the magnet has only to be lowered on to its load and the current switched on. There is nothing remarkable in the construction or form, which is simply that of a horse-shoe electro-magnet strongly cased in order to protect the coils.   For the best results the shape of the magnet should be specially designed for the work it is to be employed on.   The current is supplied by means of an armoured flexible lead connected by a plug to some adjacent wall socket; and a magnet suitable for lifting a steel billet weighing $1\frac{1}{2}$ tons, requires only about 170 watts.   There is, of course, a certain amount of risk attached to the use of lifting magnets, as they are dependent entirely upon the continuity of the supply, and further, may drop the load if it be accidentally subjected to a shock.   To overcome these difficulties magnetic lifting gear has been devised combining a mechanical safety clutch which, when the load is off the ground, closes round it, forming a cradle.   This arrange-ment is of a special nature, and is by no means usual; but the expense of its adoption is well justified where the work is regular and where it is restricted to a class of material of a shape suitable for the particular style of clutch employed. For these magnets direct current is used exclusively.

Another purpose to which the electro-magnet is put is the automatic sorting out of iron or steel filings or chips

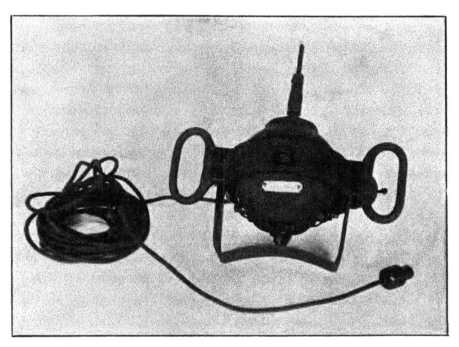

FIG. 22.— SMALL HOLMES-LUNDELL ELECTRICALLY-DRIVEN DRILL.

from a mass of non-magnetic material such as brass or workshop rubbish. These separators are of particular service in a brass foundry, where it is necessary to remove all iron from scrap brass prior to melting the latter down. They are very simple, consisting in one type of a rocking tray, from which the mixed substances fall on to a slowly rotating brass drum fitted internally with an electro-magnet. The non-magnetic material falls off the drum as it rotates into a bin, but the iron or steel is held to it until removed by a revolving brush wheel, when it drops into a separate bin. A machine for dealing with from 600 to 900 lbs. per hour takes $\frac{1}{8}$ h.-p. to drive, and consumes between 70 and 80 watts in the magnet winding. These separators are of great value in many industries, being exceedingly economical and doing better work than hand sorting at a quicker rate.

The portable electrically-driven tool has been mentioned as having produced almost a revolution in workshop methods. Taking drilling for instance: where formerly it was necessary to carry heavy work, such as plates, girders, castings, and the like to the machine, or to employ comparatively wasteful pneumatic drills with their cumbersome and leaky pipe connections, the electric hand drill can now be used, with a considerable saving in the cost of power and also enhanced convenience. The loss in transmission is almost negligible, and a light armoured flexible cable connected by a plug to the distributing main is all that is required. These drills are comparative novelties, but their working is very satisfactory, and free from many of the disadvantages of the pneumatic hand drill. They are made for both direct and alternating current, and will drill holes in iron or steel up to $1\frac{1}{8}$ in. in diameter. They may also be used for tapping, in which case a reversing switch is a necessary adjunct. The weight of a drill complete suitable for the above-sized hole is approximately 85 lbs., while that of the smallest size for drilling up to $\frac{3}{8}$ in. holes is about 20 lbs. These figures do not apply to any particular make, and are given only as a rough guide. Larger types of portable electric drills are made, the motors of which are designed for mounting on wheels or for slinging. In such cases the drill itself is operated by a flexible shaft, and may thus be used in any direction with ease. These outfits are suitable also for grinding and milling work, but special machines are made for these latter purposes.

Before leaving the subject of portable engineering tools attention should be drawn to the electric pulley block, a

FIG. 23.—PORTABLE DRILLING OUTFIT, DRIVEN THROUGH FLEXIBLE SHAFT BY HOLMES-LUNDELL MOTOR.

most useful contrivance. It is made for lifts up to 15 tons, although it is in the smaller sizes that its usefulness is most pronounced. It consists of a series wound direct current motor, fixed in a suitable frame which also contains the winding drum and the speed reduction gear. The whole is suspended as usual by a hook, or, in the larger sizes, possibly by wheels moving along a special runway, so as to serve the purpose of a travelling crane. The motor starter is also attached to the frame, and is operated from the floor level by means of a rope. This arrangement is very suitable as a cheap substitute for the travelling crane in small works or engine rooms. In its stationary form it makes a handy hoist, which may be moved from place to place as occasion requires.

CHAPTER X.

ELECTRIC POWER IN TEXTILE FACTORIES.

Comparison of British and American Practice—Comparative Efficiency of Steam and Electric Driving — Value of Constant Speed — Economy in Capital Outlay—Data of the Olympia Mills, U.S.A. —Comparative Value of Direct and Alternating Current Motors —H.-P. and Type of Motor required for various Machines— Revival of the Cottage Loom.

THE textile industries of Great Britain compare very unfavourably with those of the Continent and America so far as the adoption of electric driving is concerned. In the United States there are a large number of spinning and weaving mills operated entirely by electric power ; in this country there are, comparatively speaking, few in which electricity has even gained a footing. The cause is not far to seek. The British mills are largely of old standing, and their conditions of steam operation have been brought by experience to a point of efficiency at which in a number of cases electric power cannot offer the usual direct economies. Further, British ideas are conservative in this matter, and there exists a certain amount of prejudice which time alone can remove. On the American side conditions are different. Water-power is plentiful, being, primarily, the motive agent for an immense number of factories; and, generally speaking, water-power can only be utilised efficiently and on a large scale through the agency of the electric generator and motor. There is also less vested interest at stake, and a greater number of new mills are being put down. It will be seen, therefore, that the extended use of electric power for textile operation in America arises more out of necessity than virtue. The same necessity has never existed in Great Britain, with the result that, despite her textile supremacy, she must concede the palm to America as having originated and developed electrical power in this industry.

Although an unbiassed investigation might show that many British mills would not gain in transmission efficiency or in reduced costs by electrical driving, experience has proved that this cannot be said of all. A field for electrical power therefore exists, and it is being fully exploited with most satisfactory results by the several firms who make a speciality of the work. The crusade, however, does not stop at what may be termed inefficient mills. It is extending now in the direction of those which cannot be so classed ; and the motor is obtaining a footing in apparently unassailable quarters by reason of the indirect economies its nature renders possible. In a large number of textile operations the primary consideration is that of constancy of speed and turning moment. What are termed "broken ends" are almost the principal source of loss in cotton spinning, and these are greatly attributable to unequal speed, being produced even by comparatively small variations. In a mill driven by one main engine of the reciprocating type, there must of necessity be a certain amount of irregularity in the main drive by reason of the unequal turning moment applied to the crank shaft. This is seen in the undulation of the belt or ropes, and although the shafting and belting will damp it to a great extent, the speed of machinery in close proximity to the main drive will be more or less affected. Again, the governing of the average mill engine is not beyond reproach, with the result that when partial load is thrown on or off, momentary fluctuations occur, sometimes sufficient in amplitude to cause trouble. With the electric drive—particularly where the alternating current induction motor is employed—considerable improvement in this respect manifests itself. Voltage and speed variations of the generating plant can with care be kept within very small limits; and even though the turning moment of the generator engine be unequal it has no effect on that of the motor, which will rotate in a perfectly uniform manner.

The output of a mill naturally depends upon the speed at which the thread can be run. There are limits to this which must not be nearly approached where uneven speed is to be anticipated. If, on the other hand, inequality need not be feared—as is the case where the electric motor is installed—the machines may be run at a higher speed, and the output may therefore be increased—sometimes as much as 8 per cent.—with perfect safety. There are here two powerful arguments for the electric drive, which, if awarded fair consideration, must tell. There is, however, a still further indirect economy, namely, that brought about by reduced capital outlay. It will only apply to new mills or those in course of thorough reconstruction, but it is a fact that the installation of the electric drive in toto is cheaper than that of steam plant and shafting. Mr. Marshall Osborne, in a paper read before the Institution of Electrical Engineers, Dublin Section, in May, 1902, gave some striking figures in this respect which may well be quoted. They relate to the large 100,000-spindle Olympia

Mill put down at Columbia, South Carolina, United States of America. Estimates were prepared for electrical and mechanical power transmission, and it was found that the former would result in a 10 per cent. saving in buildings, a 61 per cent. reduction in the cost of shafting, and a 66 per cent. saving in the cost of belts and ropes. On the whole, the saving more than compensated for the capital cost of the electric plant. This is by no means an isolated example; numerous others have been published, and many show approximately the same results. These are the principal counts in the case for electric power; in addition, we have to consider minor points, such as ease of speed variation where it is necessary, better light owing to the comparative absence of shafts and belts, and more favourable all round sanitary conditions for the workers. On the whole, there is little doubt that where electric power can be generated or purchased at a cheap rate, it is preferable for textile operation than steam power.

We may now pass to a consideration of the comparative suitability of direct and alternating current motors. It is needless to reiterate characteristics which have already been fully entered into; *primâ facie*, there is little to choose between the direct current shunt wound motor and the induction motor where constant speed is concerned. The direct current motor, however, possesses a peculiarity which has not been previously mentioned because it is one of comparatively small importance in the classes of work which, so far, have been touched upon. All electrical apparatus heats up when at work, and in the case of a shunt wound motor this naturally decreases its field owing to the resistance of the winding being increased by the heat.   The result of a weakened field is that the motor runs faster, and between the beginning and the end of a day's constant run there may be a difference in speed of 5 to 10 per cent. from this cause.   The difficulty may, of course, be got over by the installation of shunt regulating resistances; but it is undesirable to give the worker power

FIG. 24.—SPINNING FRAMES DRIVEN BY POLYPHASE MOTORS.

to vary the speed of his machine which, for economic reasons, must be fixed by the management.   It is evident, therefore, that for such branches of textile work as require absolute constancy of speed the direct current motor is not very suitable; and it is now generally admitted that in this sphere the induction motor—which always tends to run at a speed synchronous with that of the generator, and is not affected by varying temperature—is the better agent.   The majority of electrically driven textile mills are operated by polyphase induction motors, although the direct current machine is much in evidence where varying speed is required.

The principal considerations governing the electric drive having now been reviewed we may pass to the more detailed particulars of the manner in which the various machines employed in textile work should be driven. In weaving, the maximum speed of operation is fixed by the strength of the yarn and also by the liability of the shuttle to "fly." It is essential in the interests of production that the speed of the loom should be as high as possible, but it is also essential that it should not at any time increase unduly, and therefore the induction motor is best suited to this class of work. Looms are driven in groups, and the power taken by each loom is approximately $\frac{1}{4}$ h.-p. In spinning, the same constancy of speed is necessary, and the induction motor is therefore again preferable. The frames, which absorb very little power, are best driven in groups from countershafting by large motors, and it is not extreme for the size of the latter to amount to 150 h.-p. The individual drive is not good practice in this class of work, and it offers no increase in economy owing to the small and comparatively inefficient motors which would be used.

Among the chief classes of machines requiring variable speed, and for which, therefore, direct current motors are the more suitable, are calico printers, calenders, beetling machines, and dye jigs. Calico printing machines require from 10 to 45 h.-p., and are, of course, best driven independently, particularly as the work is intermittent, with long stops. Compound wound motors are usually employed, and wide speed regulation must be provided for. Calenders absorb from 10 to 50 h.-p., and as they call for a heavy starting effort, individual compound wound motors are again preferable.   Speed regulation is necessary, but not to the same extent as with calico printers, a range of 2 to 1 being sufficient.   Beetling machines may either be driven independently or in groups, whichever is the most convenient. They absorb about 10 h.-p. each, and should be supplied with compound wound motors.   Dye jigs require only about $\frac{1}{2}$ h.-p., and as they call for little starting effort, the shunt wound motor may be employed, running the jigs in groups of 10 or 20.

To all the other numerous classes of machinery employed in the broad field of textile work the electric motor is easily adapted ; and in the great majority of cases the result, so far as the individual machine is concerned, will be lower working cost and increased convenience.   By the introduction of cheap electric power the cottage textile industries are being slowly revived.   The steam engine with the necessity of centralising the machinery killed them for the time being; but now that electric power is sold at a reasonable price and motors suitable for working a cottage loom may be bought for a few pounds, there are signs of revival. Large work must, of course, be carried out on the factory system, but there will always be scope for the cottage loom.

E.P.

CHAPTER XI.

ONE of the most important applications of the electric
motor is that of printing. It is a class of work where
electric driving has proved itself to be pre-eminently the
best, not only by reason of economy in power consumption
but by the extraordinary flexibility of operation it affords.
The conditions under which printing machinery works are
onerous in the extreme; steadiness of drive, extremely
wide range of speed, and instantaneous stopping in emergency being essential points. Furthermore, the load, which
at times is very variable, must not adversely affect the
operation of the press in any of the above respects. Really
good and economical results with large presses can only be
obtained with the direct current motor. The speed range
alone puts the alternating current motor out of the question
for work of any magnitude. The driving of small presses
presents no difficulty either with direct or alternating
current, but large machines of the type used for newspaper
production call for very special arrangements. The methods

employed in this class of work are interesting and ingenious, and they bring into prominence in a very striking manner the possibilities of the electric drive in the directions of flexibility and ease of control. Fig. 25 represents in diagram the main connections of the Ward-Leonard system of electrical operation. The method is a well-known one in America, and it is claimed for it that it possesses the following advantages : The press may be run up from rest to full speed in a very short time with perfect steadiness and entire absence of jerk or sudden increment of speed in any part of the range. It should be noted here that this could not be done with an ordinary starting switch, as the movement of the arm across the contacts gives rise to a

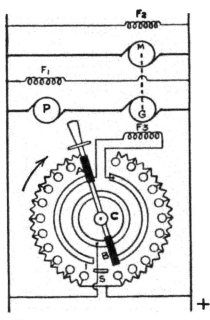

FIG. 25.—DIAGRAM OF WARD-LEONARD SYSTEM OF PRINTING PRESS OPERATION.

slight but perceptible sudden increase of speed at each contact. The press is under perfect control at all speeds, and despite wide variation of load, the speed will remain constant wherever it is set. The very lowest speed is possible with little waste, and under the worst circumstances, namely, when the press is " creeping,". the loss incurred in speed regulation will be well under 3 per cent. The machine may be instantaneously stopped in emergency from several positions by the aid of a number of pushes the pressing of any one of which automatically opens the circuit. Referring to the diagram, these results are achieved in the following manner : Assuming the pressure of the supply is 200 volts, P is the press motor wound for 400 volts and mechanically coupled to the machine through speed reduction spur gearing. M and G are respectively a motor and a generator, rigidly connected together, and wound for half the voltage of the press motor, that is to say, 200 volts. Together they form a motor-generator set of half the horse-power of the main motor. F_1, F_2 and F_8 are the field magnet windings of the three machines, and C is the controller by which the operation of the press is regulated. This latter, it will be noticed, is inserted in series with the shunt circuit of the generator G, and as it has only a˙ slight current to control, is quite small. The working of the system will be best understood by following the process of running the press motor P up to full speed from standing. The motor-generator is started up and kept running all the time at constant speed. The arm of the controller to begin with will be so situated against the stop S that the contact piece A will bridge across the resistance stop marked 1 and the left-hand semicircular contact. B, the contact piece on the other end of the arm, will bridge across the right-hand semicircular contact and the inner circular one. The field current of F_8 will there-fore flow from the + wire to stop 1, across the contact piece, through the winding, back to the circular contact, and so to the − wire. As no resistance is interposed the current will be at its highest value, and G will generate at a pressure equal to that of the mains, namely, 200 volts.

The direction in which the current is now flowing in the shunt is such as to cause the generator to oppose the pressure in the mains, and therefore, when the controller is in this position, no current will flow through P, and it will remain stationary. The switch arm is now moved in the direction indicated by the arrow, and resistance is gradually

FIG. 26.—HOLMES-CLATWORTHY PRINTING PRESS MOTOR EQUIPMENT.

inserted in the field circuit F_3. This causes G to generate at a lower pressure, with the result that current flows through its armature and also through that of the press motor P, which gradually gathers speed as resistance is inserted. When the switch arm reaches the position shown in the diagram the field current of G is cut down to a very low value, so much so that the opposing E.M.F. is nearly at zero, and the armature of P consequently has almost the full 200 volts of the mains applied to it. The next instant the field circuit of G is entirely broken, and P, which, it must be remembered, is a 400-volt motor, now runs approximately at half speed on the full 200 volts supplied from the mains. The next movement of the controller to the right again completes the circuit of F_3 through the whole of the resistance on that side; but it must be noted that the course of the flow through F_3 is now reversed, and, consequently, G will begin to generate in the opposite direction, helping the pressure in the mains and increasing that applied to the armature of P. The latter gathers speed, and as the resistance is cut out it rotates faster until it attains its maximum, which is reached when the switch arm is over to the extreme position up against the other side of the stop S from which it started. G is now generating its full 200 volts, which, being added to the pressure of the mains, results in the 400 volts required by P. This arrangement, it will be seen, gives a very wide and extremely delicate range of speed control, exactly such as is required for printing work. It is essentially simple, and there is a very desirable absence of parts liable to derangement.

Another well-known method—this time of British origin—is the Holmes-Clatworthy system exploited by Messrs. J. and H. Holmes, of Newcastle. Here again simplicity is the object aimed at and attained. In this system two motors are employed, the larger being an ordinary shunt wound machine, and the smaller a series wound motor of approximately one-tenth the horse-power of the first placed at right angles to it. The two motors are mechanically coupled together through worm gearing and an automatic

clutch actuated by a solenoid; the larger is also geared to·
the machine through spur wheels. The small motor is
used for starting the press—for which it is particularly
suitable being series wound—and also·for running it at the

FIG. 27.—SMALL ELECTRICALLY-DRIVEN PRINTING PRESS.

slow speed necessary when the paper is being led into the
machine. The larger motor is used for running the press
when it is at work in the ordinary way. Being connected
through the gear to the auxiliary motor, its armature
revolves directly the latter is started; but when current is
applied to the shunt machine it speeds up, the clutch
automatically comes out of gear, and the small motor stops.

The press can only be started by the latter, and speed regulation is obtained when in full work by aid of a variable resistance in the shunt circuit of the main motor. The operation of the press is controlled entirely by one handle, and it may be stopped instantly by the aid of emergency pushes placed in several convenient positions round the machine. The system is very largely in use and has proved itself both reliable and efficient.

In a printing office or works there are a number of machines other than presses which may be economically driven by electric motors. Binders, rulers, paper-cutting machines, paper-wetting machines, and small presses are best grouped together and driven through a short length of countershafting, as some of these require less than ¼ h.-p. each. Paper glazers, ink grinding mills, and machines calling for 2 to 3 h.-p. are in most cases best driven by independent motors. Some very useful data of the power required for various printing office machines has been prepared by Mr. A. Gibbings from actual tests. The figures are given below; but it must be borne in mind that they can only be looked upon as approximate, as much will depend upon the manner in which the motor is geared to the machine. Direct coupling is, of course, the most efficient, but belting is often necessary, and this introduces a variable factor. In addition, the viscosity of the ink in the case of presses, the size of the sheet, and the forme, all affect the horse-power required to a material extent.

LETTERPRESS, WHARFEDALE, BREMNER, ETC., CYLINDER MACHINES.

Demy, to print a sheet	24 × 18 in.	...			1	h.-p.
Double crown	,,	30 × 20 ,,	...	...	... 1	,,
Double demy	,,	36 × 24 ,,	...	...	... 1½	,,
Quad crown	,,	42 × 30 ,,	...	...	... 1½	,,

GOLDING JOBBER, MINERVA, ARAB, UNIVERSAL, ETC., PLATEN MACHINES.

Foolscap folio, 7 × 11 in. inside chase	$\frac{1}{4}$ to 1 h. p.
Crown folio, 10 × 15$\frac{1}{2}$,, ,, ,,	$\frac{1}{4}$ to 1 ,,
Demy folio, 13 × 9 ,, ,, ,,	$\frac{1}{4}$ to 1 ,,
Linotype machines	$\frac{1}{2}$,,

LITHOGRAPHIC, BREMNER, ETC., MACHINES.

Demy, to print a sheet 24 × 18 in. ...	1 h.-p.
Double demy ,, 36 × 24 ,, ...	2 ,,
Quad royal ,, 50 × 40 ,, ...	4 ,,

GUILLOTINE CUTTING MACHINES.

26 to 32 in. wide	$\frac{3}{4}$ h.-p.
38 to 44 ,, ,, ...	1 ,,

PAPER GLAZING MACHINES.

To roll 27 in. wide	3 h.-p.
,, 33 ,, ,, ...	4 ,,
,, 36 ,, ,, ...	5 ,,

INK GRINDING MILLS.

Rolls 12 × 8 in.	3 h.-p.
,, 18 × 10 ,, ...	4 ,,
,, 24 × 12 ,, ...	5 ,,

The following data of horse-power required by newspaper printing machinery are taken from an American source.

One 10-page web perfecting press 12,000 per hour...	15·39 h.-p.
,, 10 ,, ,, ,, ,, 24,000 ,, ,, ...	31 ,,
,, 12 ,, ,, ,, 12,000 ,, ,,	20·45 ,,
,, 12 ,, ,, ,, ,, 24,000 ,, ,,	29·56 ,,
,, 32 ,, ,, ,, ,, 12,000 ,, ...	28·73 ,,

CHAPTER XII.

ELECTRIC POWER AT SEA.

Power-driven Auxiliary Gear—Comparative Value of Electric and Steam-driven Auxiliaries—Direct and Alternating Currents for Marine work—Generating Plant—Standard Pressures—System of Wiring—Motor-driven Fans—Amount of Ventilation necessary Pumps—Boat Hoists—Coal, Ash, and Ammunition Hoists Capstans and Winches—Turrets—Gun Loading and Handling—Motor-operated Watertight Doors and Hatches.

IT is now some thirty years since electricity entered the sphere of marine work; at that time, and for many years after, being employed exclusively for lighting. In the early days electric power was out of the question, for the reason that the only class of motor then built was quite unsuited for the conditions which prevail at sea. In addition, the steam and hydraulic machinery employed for operating the auxiliary gear gave every satisfaction, and its wasteful nature was condoned because such machinery then played a small and comparatively unimportant part. As everyone knows, the fuel question is one of the first considerations at sea; and as power-driven auxiliaries began to multiply it was seen—especially in the case of warships—that something had to be done to keep the rapidly increasing coal consumption within bounds. The auxiliaries could not be dispensed with, and, therefore, the only course open was to find other and more efficient means for their operation than those afforded by the methods of power transmission and production then in use. The sole alternative was electric power. Steam transmission is wasteful under the best of conditions, and at the high pressures employed at sea there is always the liability of trouble with the joints. A steam pipe is easily damaged, and in the event of mishap the machine it is serving is rendered absolutely useless until such time as the injured portion is replaced. On the other hand, electrical transmission is highly efficient; indeed, the loss in the cables may be rendered almost negligible; and in the event of the wires being damaged they may be temporarily repaired, or, if necessary, a new length run, in a few minutes. The steam-driven auxiliary is essentially wasteful in itself, and not to be compared in efficiency with the electric motor. Its exhaust is liable to get frozen up in cold weather, it cannot be effectively controlled from a great distance, and its pipe connections take up much room which can be ill spared on board ship. In all these points no comparison is possible with the electric motor, which is unaffected by temperature, may be readily controlled from any number of points, however distant, and which is supplied with energy through comparatively small cables capable of being placed out of the way in any position where their presence is least objectionable. The modern motor is as reliable as any other power agent, and, owing to its characteristics, will in most cases perform its work more efficiently and expeditiously than a steam or hydraulic engine. With all the advantages on its side it is not surprising that electric power has to a great degree replaced the older systems. In the chief navies of the world it is now almost exclusively used for the operation of auxiliaries, and in passenger and cargo steamers, although there is still much to be done, it is making satisfactory progress.

When the relative qualifications of direct and polyphase currents for marine work are compared there does not appear to be much to choose between them. But since the experience gained has so far been entirely with the former, the polyphase system may be left out of consideration as actual examples of its installation cannot be quoted. The following details are not confined to British naval practice, but include that of the United States, which in some particulars is different. In both cases the standard is high, with thoroughness as the keynote of the whole design.

The generating plant now employed in the British navy consists of open type 4- or 6-pole direct current dynamos coupled direct to high speed engines of the enclosed or open class. Formerly it was customary to employ ironclad dynamos, in order to avoid magnetic disturbances, but owing to the difficulty of ventilation and the great weight, open type machines are now used, and they are found to give satisfactory results. For many years the generator pressure was 60 volts, but recently this has been increased to 100 volts with a corresponding reduction, power for power, in the weight and size of the plant and mains. The standard pressure in the American navy is 125 volts, and doubtless in both cases the more usual shore pressure of 200 to 240 volts will in time be adopted. The generators are constructed to a very rigid specification, and the plant—engine and dynamo—is highly efficient and reliable. The switchgear is in all cases of a standard description suitable for the size of the vessel. All cables and wires are lead covered, and of sufficient area to ensure a low voltage drop. They are secured to the bulkheads, etc. by brass cleats, and where it is impossible to fix them direct on the bulkhead, owing to the presence of ribs or other obstructions, they are carried on a galvanised steel plate. When the wires pass through watertight partitions or decks a special gland is used which makes the joint watertight. The main feeders run from the dynamo switchboard to heavy junction boxes, from which distribution is effected to the various parts of the ship through section and distributing boxes. No joints are allowed in the wires, all branch circuits being taken off through special boxes. The system of wiring is costly, but it is thoroughly reliable and affords convenient control.

The extent to which electric power is employed in ships of war varies according to the age of the ship and its nationality. The American navy probably provides the most inclusive examples, but the British navy is not far behind. In both cases the electric motor is largely in evidence, but the British Admiralty take no risks in this respect, and where there is the slightest chance of the motor proving unreliable or unsuitable for the work, it is not employed.

The chief auxiliaries commonly operated by electric power are: fans of various descriptions, pumps, boat, coal, ash and ammunition hoists, capstans, doors and hatches, winches, turret and gun machinery. In addition, there are sundry minor applications which possess no special features. Electrically-driven fans are employed both for forced draught and ventilation, and, as was previously pointed out, the motor is particularly suited to this class of work. The speed of rotation being high, the fan is of much smaller diameter than an equally powerful one of the steam-driven type, and thus an appreciable saving in space is effected. The efficiency of the combined set is also higher, and there is less noise and no vibration. In the American navy ventilating fans work at a pressure of one ounce per square inch, and they must be capable of exerting a 50 per cent. greater pressure at increased speed if necessary. This latter requirement is more satisfactorily met by the electric motor than by the steam engine. It involves a considerable increase over the normal speed, and this, although quite possible with a reciprocating engine, is never to its advantage. On the other hand, the speed of a motor may be greatly increased with no evil results, and, therefore, no extra attention is required or risk of breakdown incurred. The importance of ventilating machinery on board ship will be appreciated when it is stated that in certain parts, such as the engine rooms, stokeholds, and around the magazines, the air is completely changed once in every two minutes; and in the dynamo rooms, where excessive heat must be particularly guarded against, the change is effected in three-quarters of a minute. This is when the fans are working at their normal pressure of one ounce per square inch ; at increased pressure the change is, of course, quicker. The class of fan used is of the enclosed blower type, and the motor is always directly connected to its spindle. Speed variation is secured in the ordinary way by means of a regulating resistance connected in the shunt circuit.

Both plunger and centrifugal pumps are operated by motors with considerable saving in space and increase in efficiency. The shunt wound motor is used, either coupled direct or through gearing as the exigency of the case may require. The high speed centrifugal pump, like the blower, is particularly suited for the motor drive, and it is being used to an increasing extent in conjunction with small and light high speed motors.

Boat hoists require two series wound motors, one for raising the load, and the other for rotating the crane. The

FIG. 28. -- ELECTRICALLY-DRIVEN WINCH FITTED WITH AUTOMATIC BRAKE.

motors are of the totally enclosed type, and are fitted with brakes to control the load when it is being lowered. Mechanical friction brakes of various types, and also magnetic brakes, are employed. The part they play is very important, and as the slightest failure might lead to disaster, special attention is paid to their trustworthiness. Considerable experience has been obtained with motors in this class of work, and the uniform success they have achieved has led to their being specified for all modern warships. Both motors drive through spur and worm gearing, and are situated close to their work.` The space they occupy is considerably less than that which would be required by steam engines, and the convenience of having no steam and exhaust pipe lines to provide room for and keep in repair is obvious. Coal and ash hoists present no special features, and as their work is practically of the same order there is no essential difference in their equipment. Ammunition hoists are used for raising ammunition from the magazines to the decks, and vary in size and class according to the guns they serve. The electric motor is well suited to the work and is always employed. Motor driven capstans and large winches were for some time doubtfully regarded, as the work they perform is very heavy and the load extremely variable. In the warping of a ship it is possible for the strain on the cable to increase suddenly to many times its average value, in which case the motor might be brought to a standstill and either severely damaged or automatically cut out of circuit. Two methods have been employed for getting over this difficulty. In one a series wound motor of large power is used, and, by the aid of automatic switches, a resistance is inserted in series with the armature when the load becomes heavy enough to stop its rotation. This resistance allows only the maximum safe amount of current to pass, and this is sufficient to maintain a torque and keep the strain on the winch. In the second method a shunt wound motor is employed and kept continuously running at approximately constant speed. It is geared to the capstan through a friction clutch, which, when the load becomes excessive and beyond the power of the motor, slips but still maintains a strain on the winch.

The turrets of a battleship are rotated by means of shunt wound motors connected to their load through worm and spur gearing. It is good practice to insert a friction clutch in the gear train in order to protect the motor from shock due to the impact of gun firing or to sudden stoppage of the turret. For ramming home the charge in a gun a series wound motor is used in conjunction with a telescopic rammer, which it drives through chain and spur gearing and a slipping clutch. For gun handling machinery hydraulic power has been proved to possess many advantages, especially in the case of heavy guns. It is, however, being gradually replaced by the electric motor, which for some time has been exclusively used with the smaller guns. The all-round efficiency of the electric method is greater, and there is the usual saving in space gained by the absence of pipes.

The principal doors through watertight bulkheads and the larger hatches are in the modern warship best operated by electric power. In the case of doors a compound wound motor driving through worm and spur gears is used, the form of the mechanism being varied according to whether the door is of the horizontal or vertical sliding type. By an automatic contact arrangement, which comes into action when the door is fully closed, a signal lamp is lit in the "emergency station," the part of the ship from which the doors are controlled. At each door in the ship operated by this means there is placed a switch, by the movement of which the door may be opened or closed from either side on the spot. A further "limit" switch is added, which automatically cuts current off the motor when the door is fully closed, or when it meets with some obstruction. In the latter case the signal lamp would not light, which would reveal the fact that something was wrong. An improvement of the system provides that the door automatically starts to close again directly the obstacle is removed. In a case of emergency it is only necessary to press a button situated in the emergency station to close all the doors in the ship. Spring-driven gearing is set in motion which completes the circuits of all the door motors, one after the other, at intervals of about 3 seconds. By this means, without subjecting the generating plant to sudden heavy strains, it is possible to close 25 doors or armoured hatches in the short space of 75 seconds, without more than six of the motors being running at any one time. The system has only recently been brought out; but its working has proved so satisfactory in comparison with that of the older pneumatic and hydraulic gears that it is being adopted to a large extent in the American navy.

CHAPTER XIII.

ELECTRIC POWER ON CANALS.

The British Canal System—Advantages of Electric Power—Character-
istics of British and Foreign Canals—Systems of Self-Propulsion
—The Accumulator Barge—The Overhead Trolley System—
Portable Barge Motors—Disadvantages of Screw Propulsion—
Chain Haulage—Tractor Systems—The Thwaite Tractor—The
Lamb Tractor—The Tow-Path Locomotive—Polyphase Operation
—The Teltow Canal—Power Required by Barges.

As a sphere for electric power, canal haulage or barge
propulsion presents some very interesting and unique
features. It is certainly a branch of electric traction ; but
conditions are in every way so totally different to those
which obtain on land that a special class of apparatus is
required, and this must of necessity be made the subject of
a separate study. Great Britain owns one of the finest and
most comprehensive canal systems of the world. Practically
all the chief towns are linked up by waterways, which total
some 4,000 miles in length, and a century ago the greater
proportion of goods traffic was conducted over them. That
they have now fallen into almost total disuse is well known ;
indeed, the question is at the present time the subject of a
Government investigation. In the interests of economy
their restoration to something of the position they formerly
held is certainly to be desired ; and, although the outlook is
not exactly hopeful, it is within the bounds of possibility—
almost probability—that some active steps will be taken in

the near future to improve their present status. If this is to be done, electric power must be the agent. We have only to look abroad to France, Belgium, Holland, and the United States to see that electrical haulage or propulsion is by far the most economical and convenient system. Beside it horse haulage and systems of mechanical power transmission, steam, gas and oil engine self-propulsion, are all either costly, too slow, or unsuitable for the conditions which prevail in English canals. It is in this latter respect that difficulty will be met with. The Civil Engineer of a hundred years ago had no conception of any form of motive power beyond horse haulage; consequently, he designed his canals for such. He made them narrow and shallow; he did not go ahead in a straight line, but wandered here and there to avoid what were then, perhaps, insurmountable difficulties. He built his bridges over the canal very low so as to economise, and in short, evolved a waterway totally unsuited for any form of barge propulsion other than traction from the bank. To a certain extent conditions are the same on the Continent; but in the United States the main canals are comparatively modern, and they have consequently been designed in most cases for power operation.

There are numerous systems in which the electric motor may be advantageously employed; they are not, however, all suitable for British or similar waterways, for reasons that have been outlined above and which will be more fully detailed in their proper place. Primarily, the subject may be divided into two heads: methods of self-propulsion by motors fixed on the barge, and systems wherein a tractor or locomotive of some sort replaces the horse on the tow-path. The first class is capable of considerable sub-division. One of the earliest solutions to the problem was to fix the motor permanently on the barge in conjunction with a screw propeller, and to supply it with current from a battery of accumulators carried on the barge. The cells when run down were replaced at certain stations by a freshly charged set, the discharged batteries being charged up by plant installed specially for the purpose. The solitary advantage of the system lies in the fact that the barge is, as it were, a free agent and carries its own power. The disadvantages are numerous, and include the following objections: heavy capital expenditure, inefficiency, high running costs, high repairs, and waste of time in changing the sets of cells. The charging stations can only be equipped with small and comparatively wasteful plant, and, therefore, the costs of generation will be high. Even supposing power to be obtained from some local supply concern at low rates, the inefficiency of the battery alone makes the system commercially impossible. An improvement in this latter direction may be effected by the installation of an overhead wire system from which the barges take their current by a sliding contact arrangement. There still remains the objection of high capital cost; but the system is efficient, the repairs are lower, and there is little waste of time. On the other hand, the adoption of an overhead trolley wire introduces certain special difficulties which militate against the value of the system as a whole. As a rule, for financial reasons, it is only feasible to instal one overhead line, and this means a certain amount of delay when barges travelling in opposite directions meet. Further, the low elevation of many canal bridges and tunnels introduces difficulties in the installation of the line; and what is certainly objectionable, often necessitates the placing of the bare wire—possibly charged to a pressure of 500 volts—within easy reach. The type of collecting gear required is not so satisfactory in its operation as the tramway trolley pole, which is neither long nor flexible enough to meet the conditions of canal traffic. Here, it is necessary to employ the original type of trolley: that is, a little pendulum-balanced truck running on wheels on the upper side of the wire and pulled along by the barge by means of the flexible conductor that conveys the current. The apparatus is liable to get out of order and may consequently cause delay.

It will be seen that in this system capital expenditure is the chief trouble; and the larger part of it is incurred by the necessity for providing, if not every barge a large number of them, with expensive motors and switch gear. At times of loading and unloading and during slack periods of the year much of this capital expenditure is absolutely unproductive, and the finances of no canal company are such as to permit this prospect to be faced with equanimity. To get over the difficulty it has been proposed to provide a special portable motor outfit which would be placed on the barge at the beginning of its journey and transferred to another about to start back when its destination has been reached. The suggestion is alluring, and it would seem by no means difficult to carry out. It would probably be necessary to introduce a certain amount of wasteful gearing, and high efficiency could never be universally attained owing to variations in the size and shape of the barges. It is possible that more will be heard of this system, which for small and lightly worked canals would appear to be the most economical method of self-propulsion. All the above systems entail the use of a screw propeller, and this for various reasons constitutes a disadvantage. Firstly, barges as they are ordinarily built are not suited by their lines for screw propulsion; the bluff bows and the shape of the stern both contribute to inefficiency. The majority of canals—certainly all British ones—are by no means deep, and this combined with their narrowness introduces propeller inefficiency, and what is worse, results in disintegration of the banks by the wash. These objections will not apply to many of the broad foreign canals; but they are vital where British practice is concerned. A motor-driven stern wheel would be preferable in this respect, but at its best the substitute is cumbersome and more costly.

Some years ago a novel system of barge propulsion was installed on the Seine between Paris and Rouen. It is termed the Bovet chain haulage system, and from all accounts appear to be fairly efficient. Each tug barge is fitted with a series-wound direct-current motor which drives a grooved pulley made magnetic by means of an internal winding. In the bed of the river two chains are laid, one being for up and the other for down traffic. The chain is taken on board the barge and makes a three-quarter turn round the pulley, to which it adheres greatly by virtue of the magnetism. The rotation of the pulley by the motor hauls the barge along, current being supplied by an overhead trolley wire. The advantage of chain haulage lies in its suitability for tunnels and non-interference with the banks; also, it is fairly cheap to instal, and calls for comparatively little alteration to the barge. The disadvantages consist of liability of the chain to break, with consequent delay, and the difficulty that is met with at bends due to the tendency of the up and down chains to foul one another. On the whole, the system is not likely to become popular, but it is interesting by reason of its unusual nature.

We now come to the second broad method of barge propulsion, namely, that which employs some form of tractor on the bank as a substitute for the horse. It has been pointed out that so far as British conditions are concerned this is the better method. Broadly speaking, it is considerably more efficient than any form of self-propulsion, it is less liable to derangement, and cheaper to operate and to instal. There are a large number of tractor systems working both on the Continent and in America, and the results have in many cases proved very satisfactory. While most employ a form of electrical locomotive running on the towpath, there are methods in existence which provide an aerial line or rail for the tractor to run on, by this means leaving the bank entirely free from obstruction. The Thwaite system is one of the latter class, and it possesses some very good points. Two rails are provided, for the up and down traffic respectively, and are fixed at a height varying from six to ten feet from the ground to a series of heavy cast iron standards placed about thirty feet apart along the bank. The tractor consists simply of an enclosed electric motor provided with four wheels, two of which run on the top of the rail and two on the bottom. They are connected to the armature spindle by worm and spur gears, and hold the tractor—which overhangs the rails—in a horizontal position with its centre in the same plane as that of the rail. Current is picked up from an insulated conductor placed close to the running rail and is returned through the latter. The tow rope of the barge is attached to the tractor, and the motor may be controlled either from the bank or the barge, as may be desired. In the latter case, the controller is placed on the barge and connected to the tractor by flexible cables attached to the rope.

The second class of overhead tractor is exemplified by the Lamb system, which for some time was in use on the Erie Canal. In common with the electric telpher, the motor is mounted on wheels which run on a stout cable suspended from posts. It picks up its current from a special wire, and returns it through the cable upon which it runs. The system is cheap to instal, and efficient, but the lightness of its construction must of necessity render it liable to break down, with consequent delay and high cost of upkeep.

We now come to the tractor which runs on the tow-path : by far the most popular form. It is occasionally free running, that is to say, is not provided with a track ; but as a rule a narrow gauge line is provided for its accommodation. The tractor is simply an electrical locomotive consisting of one or two slow-speed series-wound motors fitted to a suitable truck and working on to the wheels through reduction gear. Over all is a protective cover, and accommodation is provided for the controller when it is not placed on the barge. In all cases current is taken from an overhead trolley system of one or more wires, according to whether direct or polyphase alternating current is employed. Of the use of the latter there are several instances in America. Between Middletown and Cincinnati some forty-two miles of canal are operated on the three-phase system by aid of tractors running on the bank. Current is generated at 4,200 volts 60 periods, and in this form is delivered to a local sub-station. A portion is transformed down to 390 volts for the operation of the adjacent line, and the remainder is converted by means of motor generators to 25 period 33,000 volt three-phase current which is transmitted to four sub-stations situated at intervals along the canal. Here, transformation to 1,170 volts is performed by stationary converters, and at this pressure the trolley wire is fed. The locomotives are provided with three-phase motors and transformers for still further reducing the pressure to the 390 volts required by the motors. The number of conversions are not conducive to economy; but even with the loss they occasion the system has been found highly economical as compared with self-propulsion in any form. On the recently opened Teltow Canal in Germany, self-propulsion, tractor, and accumulator methods have all been employed. In the former, tug barges fitted with three 20 h.-p. motors and three propellers are fed from a double overhead trolley line at 600 volts. The barges when empty progress at the rate of about eight miles an hour; and when hauling a load of 450 tons their speed is approximately three miles an hour with a power consumption of 43 kilowatts. With the accumulators in use the speed is slower, approximately 400 instead of 600 volts being applied to the motor. The type of tractor used in this instance is very complete. It is equipped with three motors, one for drawing the load, one for winding or paying out the tow rope, and the third for raising or lowering a device for varying the altitude of the rope. The use of this latter detail of the equipment is to allow other barges or craft to pass under the rope when they meet the towed train. It has been found that a load of 1,250 tons drawn at a speed of about 2·5 miles per hour requires only 16 kilowatts of energy; which, if compared with the figures previously given for self-propulsion, conclusively demonstrates the superior efficiency of the tractor system.

CHAPTER XIV.

ELECTRIC TRACTION.

First Essays in Electric Traction—Early British, German, and American Lines—Accumulator Traction—The Overhead System The Conduit System—The Surface Contact System—The Controller—Traction Motors—Truck and Car Design—Track Work—The Regenerative System—The Electrical Railway Locomotive—The Multiple Unit System—The Heilmann Locomotive—Three-Phase Electric Railways—The Single-Phase Electric Railway.

CONTRARY to the popular idea which attributes all electrical invention and progress to America, a Scotsman, Robert Davidson by name, must be credited with having taken the first practical steps in electric traction. Somewhere about 1837 he designed and built an electrical locomotive, with which he made several trips on Scottish railways between Edinburgh and Glasgow. The motor was of a crude type akin to that of the toy electric motor of the present day; but although of this class, and run only by primary batteries carried on the locomotive, it was found capable of propelling a carriage at the rate of four miles per hour. History records nothing very definite about this machine; but it appears to have been wilfully broken up by some steam engineers who resented and probably feared the entry of a possible competitor into the field of mechanical traction.

Passing over several years, during which experimental work on a small scale was steadily carried out by a few enthusiasts, we arrive at 1850, when a somewhat similar attempt to that of Davidson was made in the United States by Professor C. G. Page, of the Smithsonian Institute, Washington. He, too, built an electrical locomotive which derived its power from primary batteries, employing a motor similar in principle to the old-fashioned beam engine. Two coils of wire, through which the current was arranged to flow alternately by means of sliding contacts, acted magnetically upon a soft iron plunger, drawing it backward and forward and rotating a wheel geared to it by a connecting rod. This motor was mounted on a carriage, and despite its primitive and inefficient nature, it exerted a considerable amount of power and propelled the carriage at a fair speed. However, it shared the fate of all electric traction devices of that day. The cost of operation was excessive owing to the fact that primary batteries were used; and the necessary handling of cells delicate in their nature, and their constant replenishment with liquids and electrodes, besides being found expensive was soon proved to constitute an insurmountable objection.

With the introduction of the dynamo in or about 1864, and the motor nine years later in a form something akin to that it now bears, commercial electric traction became possible. For six years from the latter date nothing practical was achieved; but in 1879 Messrs. Siemens & Halske erected what may be termed the first practical electric railway in the Exhibition held at Berlin in that year. It was modest in size, consisting of a narrow gauge track upon which ran a miniature electric locomotive about three feet high and five feet long. It hauled three small cars—or, more properly, seats on wheels—capable of carrying six passengers each. The current was picked up from the rails; and the driver sat astride on the top of the locomotive, which was regulated from the front end. The exhibit attracted a great deal of attention, and was followed in 1881 by something more practical in the shape of a public line at Lichtenfelde, Germany. This was the first and for a short time the only commercial electric tramway in existence. But in 1883 a line was erected at Portrush, Ireland, the now notorious third rail system of supply being used. A small water turbine generating station was put down, and completed the novelty of the installation so far as Great Britain was concerned. About this time Volk inaugurated the small electric railway which is still in service at Brighton; and a conduit line, which was the first in Europe, was installed at Blackpool by Holroyd Smith. The pioneering of the commercial electric tramway must thus be put to the credit of Great Britain and Germany; but the United States now followed the lead given them principally by Siemens, and lines of many descriptions began rapidly to spring into being all over the North American continent. The number of these and similar schemes render further detailed consideration impossible; it is necessary therefore to turn from actual examples and to follow the general trend of improvement in apparatus and methods. America is undoubtedly mainly responsible for electric traction progress during the next decade. Germany contributed a share; but Great Britain practically stood still, losing the lead she originally held, and paying for it later on by being obliged for some years to go to America for all traction material. Happily this state of affairs has now ceased to exist, and British firms of the present day turn out electric tramway and railway apparatus of greater durability and equally good design as cheap as any that comes from America.

The improvements made in accumulators about this time naturally resulted in their being applied to traction. Compared with the primary battery they were a decided success; but when efficient means were discovered for supplying power through an overhead or underground conductor their wasteful nature told against them; and although still in use in certain situations which forbid the presence of either of the more efficient alternatives, accumulators are, for tramway work, practically a thing of the past.

The first overhead trolley line consisted of two copper wires—one for lead and the other for return—suspended from wooden posts by J-shaped hangers. A little four-wheeled car ran on the upper side of the two lines, being pulled along by its flexible lead as mentioned in the previous chapter. At approximately the same time the single trolley wire was introduced, the return circuit being made as in modern practice through the rails. The invention of the under-running trolley by Lieut. F. J. Sprague in 1886 marks an epoch in the history of electric traction. Failing this improvement, electricity would not possess the position it at present holds, as the over-running type presented many difficulties of operation. At first, the new system was by no means perfect or even satisfactory; and considerable modifications and improvements had to be effected before it could take its place as a commercial success. It was, however, obviously a good thing; and before many years had passed, a process of evolution gave to the under-running trolley system the position it at present holds of the best all-round method of supplying current to moving vehicles. During the transition stage from experiments to assured success the conduit system of electric traction was brought into prominence, the trouble occasioned by falling overhead wires and like failures of the trolley system offering inventors encouragement to attempt radically different means. Several lines were installed in America and a few in Europe; but although the latter met with a certain measure of success, the former, owing to faults of construction, were almost uniformly failures. Time and experience, coupled with a vast amount of ingenuity, finally succeeded in making the conduit system commercially practicable; but it was recognised from the start, as it is understood to-day, that the initial outlay as compared with that necessary for the overhead wire would be too heavy to allow of competition under similar circumstances. The conduit system has therefore always been looked upon only as an alternative to the trolley for situations where æstheticism or street difficulties are the ruling consideration.

Surface contact, or as they are sometimes termed closed conduit systems, are the product of a later date. As a class they are an attempt to secure the advantages of the ordinary conduit at the cost of the overhead wire. Many of the systems devised are extremely clever, but few have come into practical operation. Details of the merits and demerits of some of the leading methods will be found in Chap. XVII. The essential difficulties to overcome are great; but of late years much progress tending towards simplification of parts and principles has been made, and it is not beyond the bounds of possibility that the surface contact system in some form may ultimately prove the solution of many existing traction difficulties.

Adverting to improvements in detail, a most important step was made when the modern form of tramway controller was introduced, since it rendered possible a far more efficient and convenient system of control than had hitherto been used. The subject is dealt with in Chap. XVIII.; it is only necessary here to note the fact and emphasise its bearing on traction development.

Improvements in motor design have of course been many and far reaching. The modern traction motor is a wonderfully robust machine, and decidedly the most compact power agent known. It works under very arduous conditions · being subject to violent strains and exposed to dirt, damp, and many evils unknown to the industrial motor. Yet when of a good class it gives but little trouble, and provided it receives its proper share of attention is as trustworthy a machine as could be desired. In the early days the motor constituted one of the chief difficulties of electric traction. It was by no means easy to find room for it, and being then of the open type and almost entirely devoid of protection from the weather, it proved a constant source of breakdown and annoyance. However, this is all changed, and we are now in possession of a machine that has reached a stage beyond which it is difficult to see radical improvement.

The earliest cars built for electric traction were designed much upon the lines of the horse car; and by reason of the totally different conditions and the greater strain on every part due to increased speed and power, they were uniformly a source of trouble. It was not until Sprague suggested making the truck a separate structure from the car that any real progress was achieved. The idea was taken up, the motor was suspended from springs under the car, the truck was specially designed to carry it, and the electric tramcar immediately began to assume a practical shape. Since that time improvements in trucks and car bodies have followed one another fast. Better conditions for the motors have been secured by design that minimises jar and strain, pounding at rail joints has been lessened by a more flexible construction, gear losses have to a great extent been eliminated, and altogether many very great improvements tending towards efficiency, speed, and com fort have been introduced. For the time being design in this direction appears to have settled into certain definite grooves, and it is difficult to see in what way further improvement can be effected except in details.

In track work there have been great advances, it having been recognised at an early date that much of the success of a tramway depends upon the rails and methods of laying. The chemical composition of the steel is now most carefully gone into, and the whole process of making a rail is now-a-days based upon an intimate knowledge of the exacting requirements of traction work. The laying in the roadway —formerly a most cursory process, demanding nothing better than wooden sleepers or foundations—has by necessity expanded into a special branch of engineering, and the result of better design and more thorough workmanship has been reduced cost of maintenance of track and rolling stock, and far smoother running.

The latest improvement in direct current traction is what is called the Regenerative system of control. By the aid of special motors and controllers the car is enabled to return power to the line when running down hill or stopping. The system has been well tried and has given good results, showing economy in working costs, and in some respects enhanced convenience of control. Further reference to it will be made later.

The history of the electric railway proper is to a large extent bound up with that of the tramway. The earliest efforts were in the direction of electrical locomotives taking current from auxiliary conductor rails, and until a very recent date this was the only method used for the propulsion of railway carriages. Many of the early locomotives took curious shapes, particularly in America, where some were even built greatly of wood, and equipped, of course, with the huge headlight, typical of American practice. The motors used were of the open type, and they were geared to the wheels and controlled in various ways. Even with these primitive engines it soon became apparent that electricity offered many advantages over steam. The entire absence of smoke and dirt, the ease of control, and the increased acceleration at starting, were all important features, the value of which were soon acknowledged. The improvements effected in tramway motors, trucks, controllers, etc., were applied to railway locomotives with equally satisfactory results, and the early efforts of Siemens, Daft, Edison and others were gradually expanded and improved upon until the modern electrical locomotive, with its high efficiency and capability of undertaking the hardest and heaviest work, was finally evolved. At the time when it had reached a high pitch of perfection, about 1897, a new and improved method of railway operation, termed the multiple unit system, came to the fore. It dispenses with the ordinary locomotive, and equips two or more cars of the train with motors mounted on their trucks and controlled simultaneously from the driver's compartment at the head of the train. In every way the results obtained by this system are better, a greater rate of acceleration and more perfect control being secured without excessive cost. The method will be more fully dealt with in a succeeding chapter. It is now the standard system for electrical railway operation and is employed almost exclusively on all the most recent tube and suburban lines.

A very interesting development of electric traction is the Heilmann locomotive, built originally in 1895, by the Electric Traction Company, of Paris. It was practically a small generating station on wheels, the equipment consisting of an efficient type of boiler, a steam engine and dynamo, the necessary switchgear, and motors for propelling the locomotive. At first sight it would appear that the power of the engine might be more efficiently used if it were applied direct to the wheels as in the steam locomotive, but when everything is taken into consideration this is not actually so. A steam engine works most efficiently when fully loaded, but in steam railway operation for a great part of the time the engine of the locomotive is not running at its maxium output, and therefore works at reduced efficiency. Under the Heilmann system the engine could be run at full load irrespective of grades, and a method of control was used which varied the current delivered by the generator in proportion to the pull on the locomotive, and the voltage in proportion to the speed. There was no waste in starting or regulating resistances, small friction losses, and few working parts. It was claimed that a 1,500 h.-p. Heilmann locomotive, costing £6,000, would be able to haul a 50 per cent. greater load at 50 per cent. higher speed than a steam engine of similar weight and horse-power. The idea occasionally recurs in some form or another, but nothing is now heard of the Heilmann locomotive, due, no doubt, to the introduction of the multiple unit system in conjunction with a third rail or overhead conductor for supplying the current.

The ease and efficiency with which alternating currents may be transmitted and transformed naturally led many to consider their adaptation to railway traction. Messrs. Brown Boveri, of Baden, Switzerland, appear to have been the first to utilise polyphase currents for this purpose. In 1895 they installed a three-phase electric railway at Lugano, Switzerland, using induction motors, and a system of picking up the current consisting of two overhead wires for two legs of the circuit, and the rails for the third. The line was successful, and was soon followed by others erected by the same enterprising firm. It has been pointed out that the induction motor does not make a good variable speed machine, and for this reason it is in a measure unsuitable for traction work. It was deemed, however, that the advantages in transmission afforded by alternating current more than outweighed this trouble in many cases, and hence the installation on the continent of a number of railways operating on the three-phase overhead wire system. There is certainly much to commend in this method, and until a very recent date it was looked upon as the only commercially possible way of applying electricity to main line railways. Within the last few years, however, another aspect has been put on the situation by the introduction of a class of motor which in every way works efficiently and reliably with single phase alternating current. Its control is more economical, and as easy as that of the direct current motor; it is capable of widely varying speed, and requires only one overhead line. The importance of this development cannot be over estimated, as it brings us within measurable distance of a really practicable means of adapting electricity to the stringent requirements of main line service. The general principles of the system are gone into in Chapter XXIII.; here it is only necessary to record the almost unanimous verdict of engineers, that if main lines are to be electrified the single phase alternating current system is the one that must be adopted.

CHAPTER XV.

THE overhead wires of a trolley system may be suspended in three ways—(1) by means of span wires fixed to posts or to the walls of houses; (2) on the arms of poles placed in the centre of the road; (3) by means of poles situated at the side of the road and provided with long bracket arms. The three systems are shown in diagram in Fig. 29. Span wire construction on the whole gives the best results, and is applicable to practically all conditions. Where poles have to be provided, as in the figure, it is the most expensive system to instal, but if the objections of property owners can be overcome, the span wires may be attached to the walls of buildings by aid of special rosettes, thereby considerably reducing the cost. A seven stranded galvanized steel cable, generally composed of No. 14 wires, is used for suspending the trolley wires. It is fixed to each of the poles or rosettes by an insulating turnbuckle, which allows slack to be taken up, and the strain on the wire to be rightly adjusted. The trolley wire is suspended by an

insulated hanger or " suspension," a typical form of which
is shown in section in Fig. 30.   The span wire passes in a
semi-circle round the body of the hanger, which is kept in

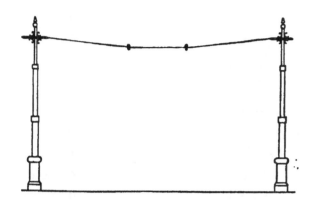

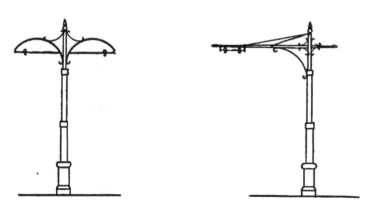

FIG. 29.—SPAN WIRE, CENTRE POLE, AND SIDE POLE
OVERHEAD CONSTRUCTION.

position by the two extended grooved lugs, through which
the wire also runs.   The body of the suspension is usually
made of bronze so as to be free from the effects of corrosion, and it is mounted on a screwed steel bolt, as shown,

the space between the two being filled with a suitable insulating material. To the bottom of the bolt, and at right angles to the span wire, is attached the " ear," a long clamp that partly or wholly encircles the trolley wire, forming the means of support. For straight runs a usual length for this clamp is fifteen inches, and it is fixed to the trolley wire either mechanically or by soldering, according to the design. The mechanical ear clamps the trolley wire by the agency of screws, or some other tightening device, and it is therefore both easy and cheap to fix. On the other hand, it often gives rise to sparking when the trolley wheel passes over it, and owing to its necessary rigidity there is increased risk of the conductor breaking at the ends of the ear by reason of the bending occasioned by the passing trolley taking place entirely at these points, instead of being

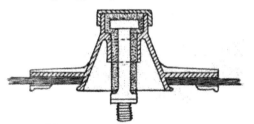

FIG. 30.—SECTION THROUGH HANGER.

distributed over the length of the ear as in the lighter soldered variety. These latter take longer to fix, and in this respect are more costly; their maintenance, however, and that of the over-head wire, is found to be less than where mechanical ears are employed.

Some years ago it was customary to use silicon-bronze for the trolley wire, owing to its strength. This has now been superseded almost everywhere by hard drawn copper wire, which although of lower tensile strength, has a much higher conductivity and is easier to handle. A wire of circular section is now almost universally used, but there are other shapes on the market. Chief among these is what is known as "figure 8" wire, the object of this and similar forms being to permit of the ear being attached to the wire so that it in no way interferes with the running of the trolley wheel. The idea is good theoretically, but in practice grooved wire of any sort is found to give trouble owing to its tendency to twist, and the difficulty experienced in handling it. By Board of Trade regulation it is necessary in this country to split up the overhead wire into at least half mile sections, and for this reason the wire is usually made in half mile lengths. Jointing is always undesirable, as it is necessary to employ a sleeve, which materially increases the diameter of the wire and causes sparking. The joint may, of course, be made at an ear, but the fact that the work is carried out *in situ* under somewhat unfavourable conditions is not conducive to a good job. The splitting up of the wire into sections, fed individually by tappings from the feeder cables, necessitates the introduction of a device at the points where it is split, which, while allowing the trolley wheel to run smoothly, will not permit of one section being made alive from the other during the passage of the wheel. Such a device is termed a "Section Insulator," and several different types of various merits may be obtained. The section insulator is really two ears mechanically fixed together, but insulated from one another, and into these ears the ends of the trolley wire are fastened some distance apart. In one form the space between the two ends is filled in with a piece of hard wood, which allows the wheel to pass with facility, but effectively insulates the wires from one another. In what is known as the divided arc type the space between the two wires is filled with small segments of metal, insulated from one another, but presenting a smooth passage to the wheel.   The arc which occurs at the moment of breaking is thus split up between the segments, and dies out before it can bridge across to the other wire.   In the former type the arc is not split up, but owing to the length of the break it cannot follow the wheel from wire to wire.

In places where telephone or telegraph wires cross the overhead line it is usual to provide what are termed " guard wires " to prevent contact with the trolley wire in the event of the former breaking or falling.   They are hung about two feet above the line, and are carefully connected to earth at intervals.   Galvanized steel, or in special cases hard drawn copper wire, is used, the area being sufficient to carry a current 50 per cent. greater than that required to open the circuit breaker of the feeder.   Guard wires are found to be in themselves a source of trouble through breaking, and although insisted on by the Board of Trade are not popular with traction engineers.

In an overhead equipment there are a very large number of fittings such as special insulators, crossings, frogs, etc., which from considerations of space and scope cannot, here, be dealt with in detail.   The term " crossing " speaks for itself, and the fitting known by this name is simply an arrangement installed to provide a path for the trolley wheel where two overhead lines cross one another.   The wires are either clamped or soldered into it, and the two limbs of the crossing may be insulated from one another or not as occasion requires.   It is preferable that they should not be insulated, as the special form of crossing necessitated is often a source of trouble.   A " frog " is to all intents and purposes the equivalent of a railway point.   It is employed at places where one or more tracks branch off, and either automatically or by hand operation guides the trolley wheel on to the proper wire.   This fitting is attached to the line in the same way as the crossing, and in both cases very careful adjustment is necessary if smooth working is to be secured.

The centre pole system of overhead construction has the advantage of being easily adaptable to artistic requirements, and at the same time it is good mechanically.   In these respects it is preferable to the side pole system, which often calls for an extremely long, weak, and somewhat unsightly bracket arm.   Neither of these methods can compare in smoothness of working with the span wire, which by its very nature is flexible and non-rigid.   In early centre and side pole construction it was customary to attach the trolley wire to the bracket arm by means of a rigid insulator, but this was soon found to be unsatisfactory.   In modern practice, what is termed a "flexible suspension" is employed, and the results are very much better.   On referring to Fig. 29 it will be seen that the trolley wire is suspended from an insulator mounted in a short piece of wire attached at its ends either to special fixtures or to some suitable parts of the bracket itself.   The result of this is to leave the wire free to move up and down at the point of suspension, thereby securing smooth working and a considerably lower repairs account.   The fittings used in these two systems are in the majority of cases similar to those described above.

A wooden pole line, in this country, is rare; but it is common practice in American suburban districts.   The woods most suited for the purpose are hard pine, cedar, and chestnut; and the poles should be free from splits and large knots and of as straight a grain as possible. The modern steel pole is the outcome of the necessity for stronger and more ornamental construction. It is usually built up of four parts, a decorative base, and three lengths of solid drawn or lap welded steel pipe of different diameters. In the fitting of these three sections together the larger pipe in each case is brought to a red heat, the smaller one slipped into it, and the joint passed through a special roll which draws it out smooth and welds the two together. In the process the placing of seams in a line is carefully avoided, as this would tend to weaken the pole. It is customary under ordinary circumstances to sink the pole a distance of five or six feet into the ground, the bottom resting on a concrete flag. The base is then surrounded with about six inches of concrete, well rammed, and a week at least is allowed for this to set before any strain is put upon the pole. The operation of planting the poles is performed by the aid of a special tower wagon, with the use of which a small gang of men can, under favourable circumstances, plant as many as forty poles a day. Sheer-legs are also employed, but they do not give the same convenience as the tower wagon and are not so portable. When the line is erected the strain on the pole causes it to bend considerably, and it is therefore necessary in planting to rake the pole a few inches backward, so that it may assume an approximately straight position when the strain is put upon it. The amount of this rake will vary considerably with the quality of the soil and the weight and strain the pole will be called upon to support. Under the best circumstances it will be about three inches, but in exceptional cases it may be necessary to allow as much as 18 or 24 inches rake. In centre pole construction no rake need be given on a straight run, as the two sides balance one another.

The importance of good track work was referred to in the previous chapter. Badly laid or designed track must inevitably result in high cost of upkeep, not only of the rails and their foundation, but also of the rolling stock, motors, and gears. Since its inception some two hundred years ago, the tramway rail has passed through many phases. Originating as a plain track of heavy wooden planks, laid in the ground, it gradually developed, becoming heavier and stronger as the weight and speed of rolling stock increased. The first step was the attach ment to the original wooden track of cast iron strips some five or six feet in length. These were spiked down, and while they lasted were naturally an improvement so far as reduced tractive effort and smooth running were concerned. The fragility of the strips and the large number of joints were found to give trouble ; hence the next step was to substitute wrought iron strips of greater length, but still attached to the wooden bed in the same primitive way. Gradually, the flat strip was improved upon ; it took a grooved shape, and instead of being only laid on the wooden track it was extended so as to cover the entire upper surface and part of the sides. Thus it was made more secure, and the whole track became more durable. However, considerable difficulties still existed, only to be overcome by the introduction in 1887 of the

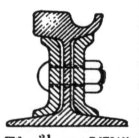

FIG. 31. — SECTION THROUGH TRAMWAY RAIL AND FISH-PLATES.

rolled steel girder rail, the type in use in improved forms at the present day.   Fig. 31 shows a standard British girder rail in section.   They are laid in 30, 45, and 60 feet lengths, the latter being preferable except from the point of view of handling, as it minimises the number of joints. The size of a rail is specified in pounds weight per yard of length; and it varies according to the class of traffic and size of rolling stock with which it is to be employed.   The Manchester Corporation use a fairly heavy rail of 98 lbs. per yard; the Dublin United Tramways a rail of 92 lbs.; the Glasgow Corporation employ a rail of 84 lbs. weight,

and the Bristol and Kidderminster Tram-ways—both early lines —are equipped with 76 and 74 lb. rails respec-tively.   The tendency of up-to-date practice is

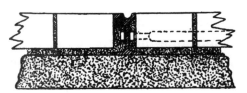

FIG. 32.—TRACK CONSTRUCTION, SHOW-ING BEDDING AND RAIL.

to increase the weight of rails, as it has been found that the reduced cost of upkeep warrants the extra capital expendi-ture.   Fig. 32 shows the manner in which tramway track is usually constructed in this country.   A bed of concrete at least six inches thick forms the foundation, and upon this the rails are laid direct, a further layer of Portland cement concrete being added, as shown, to cover the foot of the rail.   Upon this the paving blocks are placed, their cementing and binding together being carried out in various ways according to the class of paving.   At intervals of from 6 to 12 feet tie bars, consisting of steel rods screwed and nutted at each end, are installed to preserve the gauge of the track.   The question of gauge is a very important one, if only in the interests of standardisation. What is termed the standard gauge of Great Britain is a distance apart of 4 feet 8½ inches measured from the inside edge of the rail head. This appears gradually to be becoming universal, as it has been found to give the right proportions to the car and to allow convenient room for the motors. A narrower gauge is necessary at places where streets are restricted in width; but the cars are comparatively cramped, and difficulty is often experienced in fixing the motors and gears.

To return to the question of track construction. At each joint the rails are bound together by two fish plates, which are shown in end view in Fig. 31. They are attached by six or eight one-inch steel bolts, and their purpose is to assist in obtaining a rigid joint. Various forms are used, but the section shown is probably the most common. In addition, a sole plate or anchor joint is attached to the bottom flange of the rail, thus further restricting any relative movement between the two ends. The sole plate—as its name implies—is simply a plate of steel secured to the flanges by bolts and nuts. The anchor joint—mostly employed in England—consists of a piece of ordinary rail about two feet long attached in an inverted position by bolts to the bottom of the track rails at the joint and embedded in the concrete foundation. The latter makes the most solid construction, as it not only binds the rails together but anchors them securely at the point they particularly need support. A modern system, at present not in much use in this country, is the Falk Cast Welded Joint. It entirely abolishes fish and sole plates, and makes the track electrically continuous throughout its length in a manner impossible to attain with the ordinary plates. In making the joint, the two rails are butted together and a special mixture of molten cast iron is run into a mould surrounding the joint. The result is that both the rails and the metal become welded together, forming a most efficient joint in every way.

Points, and special work at crossings and turn-outs, have of late years been greatly improved. The latter work is provided with cast steel renewable pieces, as the wear is heavy at the crossing point. Movable points are generally hand operated by means of levers fixed on the kerb; automatic points are made, but their use is not desirable.

As before mentioned, the return circuit, except in the case of the conduit system, is always made through the rails. A double trolley was tried and found unworkable; but the track, if properly constructed, serves the purpose excellently. In early days it was anticipated that fish and sole plates would provide sufficient electrical connection between the rails; but experience soon proved that, owing to the formation of oxide of iron between the two surfaces, the conductivity of the track fell to a very low value. The result of this was a heavy loss of power in the return circuit, and the shunting of much of the current to earth, where it would flow mainly along gas and water pipes, causing considerable trouble through electrolytic corrosion. The remedy was obviously to connect or bond the rails together by some more efficient means, and this was attempted by riveting a piece of iron or copper wire to the two rails at the joint. Although an improvement for a time, such bonds were never satisfactory. The area of contact with the rail was too small, and rusting soon occurred, with consequent deterioration of the connection. Improvements

E.P.   M

followed directly it became clear that a good bond was
not only a desirability but an absolute necessity; and
at the present day there are a very large number of types,
many of which are excellent, and all an advance on the
early primitive method.   It is impossible to detail even a

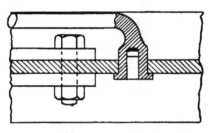

FIG. 33.—CHICAGO BOND.

tithe of the varieties, but four
typical examples may be
taken.   The Chicago bond
consists of a copper rod or
flexible cable provided at each
end with a split thimble bent
at right angles to it.   These
thimbles are inserted in holes
drilled in the web of the rails, and steel drift pins are
driven home into the thimbles from the side of the rail
opposite to that on which the bond is situated.   By this
means the copper thimbles are expanded until they make
thorough contact in every part with the rail.   The projecting ends of the thimbles are flattened over so as to prevent
the bond from drawing out of the hole.   This bond—the

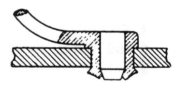

FIG. 34.   CHICAGO CROWN
BOND.

essential part of which is shown
in Fig. 33—is placed outside the
fish plate, bridging it, and is
therefore of a considerable length.
It is efficient as a bond, but it is
open to the objection that to fix
it is necessary to have access to each side of the rail.
For this reason the Chicago Crown bond is perhaps
a better type.   In one form it is much shorter, and
is placed under the fish plate, being composed of a flattened bunch of copper wires welded into the expanding

terminals. In this type a steel pin is also used for securing the necessary contact, but it is driven from the same side of the rail as that on which the bond is fixed, the shape of the thimble being varied specially to allow of this. The driving home of the pin, besides expanding the terminal in the hole, forces out its end, which is flanged internally, and the bond when fixed takes the shape shown in section in Fig. 34. The terminal of a third type of bond named the Columbia is shown in Fig. 35. A conical thimble is inserted in the hole in the rail; and the bond, which also has a conical head, is slipped into this from the other side. A hand press is then applied, and the terminal is expanded until it makes a tight fit. With this class of bond it is of course necessary to have access to both sides of the rail. A totally different type from any of the foregoing is the plastic

FIG. 35.—COLUM-BIA BOND.

bond, for use under the fish plate. It consists of a plug of plastic-conducting amalgam encased circumferentially by a flat piece of cork. At the place where the bond is to be applied the web of the rail and the fish plate are carefully scoured and amalgamed. The latter is screwed up so as to compress the bond, and the current passes from one rail to the other through the fish plate and the two bonds placed one at each end. When carefully installed these bonds have met with considerable success, and they possess the advantages of cheapness and easy application. Bonding in Great Britain has been very carefully gone into for the reason that the Board of Trade requirements in the direction of pressure drop in the rails are stringent. To avoid the evils consequent upon the return current leaving the rails and flowing through the earth and neighbouring pipes, it is specified in the rules of the Board of Trade that the pressure difference between the nearest and farthest points of the rails to and from the power station shall never exceed seven volts.   In all traction stations this important detail of working is checked by a voltmeter permanently connected between the negative, bus bar and the farthest point of the track.   It is also provided in the regulations that only the negative pole of the system shall be connected to the rails—never the positive—as this considerably minimises electrolytic action.

CHAPTER XVI.

THE CONDUIT SYSTEM.

Origin of the Conduit System—Favourable Points—Disadvantages—
The Bentley-Knight Conduit—The London County Council Con-
duit—Construction of Track—Details of Conduit—Insulators—
Drainage—The Collecting Plough—The Side Slot Conduit—
Difficulty with Points and Remedy—The Bournemouth Conduit.

STREET conduit electric traction has, at the present time,
reached practically the highest pitch of perfection possible
with existing methods. Improvements in detail may no
doubt be anticipated as more experience is gained, but so
far as fundamental methods go, there is every indication
that finality has been reached. The system owed its rise
to the extremely inefficient methods of overhead construction which obtained in the early days of electric traction.
It was from the beginning fully recognised that it would
be costly in comparison with the trolley, and that difficulties
would be certain to arise in its practical working which the
latter—owing to its flexibility and handiness—would be
without. However, aesthetic objections, coupled with fear
of accident from falling wires, temporarily gained the day
for the conduit system. It came into being in primitive
form both on the Continent and in America, and although
at the present day the trolley is undoubtedly the better
system as regards cost and ease of working, the conduit is
in considerable use in places where aestheticism prevails,

or where conditions such as obtain in some of the London streets render overhead wires an impossibility. The chief points in favour of the conduit system are: No unsightly overhead wires, and consequent freedom from the risks attendant upon their use; and an insulated return circuit such as is commercially impossible with the trolley. As is well known, the first is in many cases an important consideration. Trolley wires far from add to any beauty a town may possess, and although in many instances this is a negligible quantity, the march of progress has still left us sundry works of Nature and the old-time architect which even the most prejudiced utilitarian would be loth to see marred by the handiwork of the overhead constructor. The second point in favour of the system—its insulated return—is of importance, as thereby all troubles from electrolysis of underground pipes are obviated, and the necessity for bonding the track is done away with. Some very early systems used the rails as the return circuit, but it soon became apparent that the above advantages were cheaply bought by the extra expense of installing two conductors instead of one in the conduit, and all modern work is carried out upon these lines.

Turning now to the other side, there are some important disadvantages to consider. The initial expense of the conduit system is very great—at least double that of the trolley—and it certainly does not afford a commensurate amount of convenience or reliability in working except in very special cases. Again, road obstructions during the laying of a conduit track are considerable; much more than in the case of the overhead wire, and in crowded thoroughfares this is a matter for grave consideration, as it

may adversely influence the trade of a whole district for
several weeks. The engineering disadvantages of the
system vary greatly. They depend chiefly upon the type
of conduit, the materials used in its construction, methods
of supporting and insulating the conductors, and the
arrangement of the track at points and crossings. According to the amount of forethought and experience brought to
bear upon each identical case, so these disadvantages may
either be very real or negligible. Taking an early example,
Fig. 36 represents the Bentley-Knight conduit, installed in
Cleveland, Ohio, in 1883. It was good for those days, but

FIG. 36.—BENTLEY-KNIGHT CONDUIT CONSTRUCTION
IN 1883.

in the light of modern practice contains almost every fault
that a conduit system can possess. Putting aside the
primitive construction of the track, the conduit itself is
weak in the extreme; the delicate iron trough enclosed in
timber would never withstand the strain of traffic, and it is
incapable of being tied efficiently to the rails. This latter
point is in all conduit systems one of the greatest importance. According to the season of the year and the temperature, the road-bed expands and contracts, and as the
conduit is the weakest point, so it will automatically take
up these variations unless it is rigidly prevented from doing
so. The consequence is in hot weather the slot, unless
properly tied, will close up, preventing the passage of the
collecting plough. In like manner, heavy traffic tends to

close the slot, and, unless it is very strongly constructed, to break up its edges and generally incapacitate it. The conduit in Fig. 36 is also too small to allow of efficient drainage; indeed it is doubtful whether drainage of any sort could be applied to such a system, and for this reason, and owing to the difficulty of obtaining efficient insulation, it would surely fail in the long run.

Perhaps the best example of street conduit work is that installed on the tramway lines of the London County

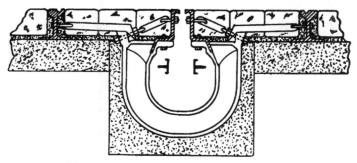

FIG. 37.—SECTION OF L.C.C. CONDUIT AT YOKE.

Council, and an outline of its details should give a very fair insight into conduit practice. As usual, a foundation of concrete forms the base of the track, and the conduit itself is built up of concrete. The most important parts of the system are the yokes, for upon their strength and rigidity depend the successful working of the line. Fig. 37 shows a section of the track at a yoke, and illustrates the manner in which the whole arrangement of rails, slot rails, and yokes are tied together. The latter are castings designed to secure great strength in order to prevent the slot closing up under a heavy weight on the surface of the road. They are spaced at a distance of 3 feet 9 inches along the track, to the rails of which they are tied by bars in the manner shown. The slot rails, consisting of Z shaped steel bars, are bolted to the crown of the yokes, and tied to the same lugs as the rail tie bars by similar but shorter rods. They are 7 inches high; just sufficient to take the stone blocks with which the road is paved. The setting is arranged to allow of a three-quarter inch slot, a width very usual in this country, but often exceeded abroad, where gaps up to $1\frac{1}{4}$ inches have been used. The slot rails are not secured together by fish plates, as their fixing to the yokes renders this unnecessary. The lip on the lower side of the rail head is provided so that water flowing into the slot may be dropped off into the centre of the conduit clear of the insulators, instead of trickling down the sides. The con-

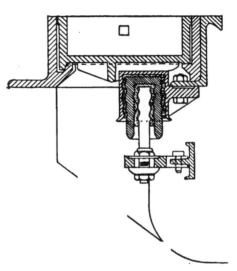

FIG. 38.—DETAILS OF METHOD OF SUPPORTING CONDUCTOR RAILS. L. C. C. CONDUIT.

duit itself is built up by means of sheet iron formers, round which the concrete is packed. The formers are placed inside the yokes, which they just fit, and after the concrete is set they are withdrawn. The width of the London County Council conduit is $14\frac{1}{2}$ inches, and its height from the bottom to the lower edge of the slot rails is 17 inches. This is sufficient to take the two conductor bars comfortably, and to leave ample space beneath them for drainage and the temporary accumulation of dirt. The conductors are of mild steel, T shaped. In some early lines copper rods were used, but although the conductivity of steel is comparatively low, its long wearing properties are valuable, and in any case the size of angle iron necessary to give the requisite strength and area of contact is such as to keep the loss through resistance within quite reasonable limits. The conductor rails are made in 30 feet lengths, and are bonded together. They are supported in the manner shown in Fig. 88 by insulators, one of which is placed at the centre of the rail, and others at each of the joints. These insulators are of high importance, and as the welfare of the system depends greatly upon them, it is necessary that they should be efficient in every way. The type used by the London County Council is sound both mechanically and electrically. The clip which supports the conductor rail is bolted to the end of the corrugated steel central pin, which is cemented into a porcelain insulator. The latter is in turn cemented into a cast iron box, which is bolted to the bottom of the slot rail as shown. At the points where the insulators are placed, the conduit is enlarged horizontally in a manner which allows of easy inspection or repair. As will be seen, a removable cover is provided in the roadway over each insulator. Drainage of the London County Council conduit is effected by sumps, or pits, placed under the roadway about 40 yards apart, and connected to the conduits and sewers by 12 inch pipes having a pronounced fall. The sumps measure 5 feet by 8 feet internally, and are provided with a man-hole giving access from the roadway for cleaning purposes. At approximately every half mile the conductor bars are sectionalised, a break of about 2 feet being allowed.

A most essential, but unfortunately very weak, link in the chain is the plough, the device used in the conduit system for collecting and returning the current from and to the conductor rails. The difficulties met with in the design of ploughs will be appreciated if the conditions under which they work are considered. The slot is only three-quarters of an inch wide, and therefore the plough must not exceed one-half inch in thickness. In this restricted space two heavy conductors, at a difference of potential of 500 volts, have to be placed, and it is necessary that they should be thoroughly insulated from one another and from earth, and also that they should be mechanically protected from injury. In addition, the plough must be strong enough to perform its work, and to clear by force any small obstacle jammed in the slot; but, on the other hand, it must not be too strongly constructed, as in the event of serious mishap it is preferable for the plough to break than for the truck frame or track to be injured. The collecting gear or shoes must be so proportioned as to allow of ample contact surface, and the maintenance of an efficient connection under all circumstances. The London County Council plough meets these somewhat arduous conditions well. The shank consists of two thin steel plates joined together by two thicker plates, through which the flat conductors are run. These latter are insulated with vulcanised rubber and tape, and the whole structure is secured together by countersunk screws. This is the portion which passes through the slot; it is capped by a head of gunmetal bolted on to it, and attached to the plough-carrying device on the car. The lower end of the plough is of impregnated maple, fixed to the shank by means of plates and bolts.

It carries the cast-iron collecting shoes, which are pressed against the conductor bars with a pressure of about 3 lbs. by springs acting upon the shoes through links. The plough is hung to the truck of the car by means of two bars fixed across the frame horizontally. Lugs on its head ride upon these, and the plough is free to move laterally the entire width of the car, thus securing the necessary flexibility. In the event of an accident, such as the plough taking the wrong path at a junction, no damage should be done, as it would simply slide along the bars and fall off at the ends, which are left open and unobstructed for this purpose. Any tendency of the plough to lift is counteracted by a further set of bars that provide an upper bearing for the lugs upon which it hangs.

The conduit system as usually laid down is constructed with a slot placed centrally between the two rails of each track. There is, however, another method, used in several places, in which the conduit is situated under one of the track rails, the slot taking the place of the groove in the rail. The advantages of the system are that the appearance of the road is improved by the absence of the special slot rail, and that ordinary vehicular traffic is less interfered with owing to the comparatively unbroken surface. The disadvantages lie entirely upon the engineering side, where a number of special difficulties have to be faced. The ordinary form of slot rail will not suit, as the weight of the car bears on it; and, therefore, a special split rail must be employed, which at times presents difficulty in its efficient installation. The fact that the plough runs in the same line as the wheels means that it is plentifully bespattered with mud and water in wet weather; and this is far from good for the piece of apparatus that is admittedly the weakest point in the conduit system. The great objection to the side slot lies, however, in the difficulties experienced at junctions with the points. So far as the plough goes, a point that will guide it into the desired path can be easily constructed ; but it must be remembered that the same point must suffice for the wheels, and this, owing to the weight of the car, is quite another problem. Obviously, the point must be efficiently supported to bear the pressure, and it must be longer and more rigid than one which would suffice

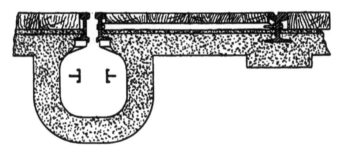

FIG. 89.—SECTION OF BOURNEMOUTH SIDE SLOT CONDUIT.

for deflecting the plough alone. The difficulty is overcome in a very simple and effective manner by reverting to the central slot system at junctions. The wheels continue in the track, which, for that portion, is built solid in the ordinary way ; but the plough is deflected from the running rail slot to one laid down the centre of the track, the points being operated together as in the central slot system. To attain this end it is, of course, necessary that the plough should be movable horizontally in the same manner as in the L. C. C. tramways. The construction of a side slot system is shown in Fig. 39, which is a section of the track between the yokes of the Bournemouth tramways. The inner or "off side" rail is the slotted one, which is always preferable as it minimises the amount of roadwork by reason of the two conduits being close together. Further, the openings are then at the highest part of the road, which means that water is less likely to flow into them. It will be seen that the slotted rail consists of two separate rails placed close together. The outer one of the two is that upon which the tread of the wheel bears, and both are supported upon and bolted to substantial yokes placed 3 feet 9 inches apart. The inner rail is tied to the opposite side of the track by the usual bars, and at the yokes both rails are tied to extended lugs in the ordinary way by short tie rods. In details of its construction the Bournemouth conduit is very similar to that of the L. C. C. Extended pits are provided at each of the insulators, which are of the suspended pattern, but road manholes to give access to the boxes are not used. This, of course, necessitates pulling up the road when it is required to get at the insulators, but, as such occasions are rare, it need not be counted a disadvantage.

CHAPTER XVII.

THE SURFACE CONTACT SYSTEM.

Principle of the System—Magnetic Operation of the Switches—Lineff System—Diatto System—Dolter System—Electrical Operation of the Switches—Wynne System—Johnson-Lundell System—Mechanical Operation of the Switches—Allen and Peard System—Kingsland System—Summary of Electrical, Magnetic, and Mechanical Systems.

In a previous chapter the surface contact system of electric traction was characterised as an attempt to secure the advantages of the conduit at the cost of the trolley. That it offers an alluring scope for the exercise of the inventive faculty is proved by the numerous and diversified methods which have from time to time been proposed and experimented with. The records of the patent office teem with such; and yet, despite the apparent simplicity of the problem, and the determined efforts that have been made to solve it, only a very limited number of attempts have met with even a measure of success. Broadly, the main idea is a follows. A bare conductor or third rail placed on the surface of the road is, of course, impossible; but, it is argued, if that conductor be split up in short lengths, or, preferably, if a row of studs be substituted for it, and only such portions of the conductor or such studs as the car covers at any one time be made alive, the advantages of the third rail system are secured, and the risks attendant upon its use entirely obviated. Necessarily, some system of

automatic switching, dependent for its action upon the
presence of the car, must be employed to switch current on
to the contacts when the car is over them and to cut it off
when the car has passed; and if such switching can be
carried out with perfect reliability, the proposition may—
engineeringly—be considered sound.   This is the crux of the
matter, and the pitfall into which the majority of inventors
have stumbled.   Their switches—often ingenious—have not
proved reliable; and hence many a system which, perhaps,
embodied some admirable features, has failed as a whole
because the very essen-
tial — the automatic
switch — has not in
practice justified the
hopes built upon it.
It is not to be sup-
posed that the surface

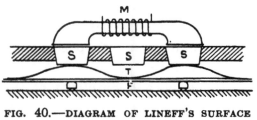

FIG. 40.—DIAGRAM OF LINEFF'S SURFACE
CONTACT SYSTEM.

contact system has uniformly proved a failure, or that
it has no future before it.   There are several lines, both
in this country and abroad, which are working regularly,
and, so far as one can hear, satisfactorily.   No epoch-
making system, it is true, has yet been devised; but it
must be recognised that nowadays far greater and
apparently more impossible problems are being solved, and
hence it is unsafe to prophesy complete lack of success,
even of a method against which so much has been said as
surface contact traction.

The system may be broadly divided into three heads
according to the manner in which the car acts on the
switches—that is to say, magnetic, electric, and mechanical.
In each section a few representative examples have been

taken, and the descriptions are chiefly confined to such systems as have seen practical service to some extent.

This is not the case with the method shown in Fig. 40, which is introduced with a view to demonstrating the weakness of some of the early suggestions. It was patented by Lineff, in 1888, and is one of the first instances of the use of studs in place of the sectional rail. The iron studs S S S were to be placed very close together along the centre of the track, and in a channel underneath them was to be the feeder F of bare copper strip supported upon insulators. On the upper surface of this was to lie a thin steel tape T, and under the car was to be fixed an electro-magnet M, in length practically equal to three studs and the spaces between them. Fixed adjacent to the magnet were to be the usual collecting brushes or skate. The action was simple. When the magnet arrived centrally over the studs it would magnetise them, and they in turn would pick up the steel tape in two places, thus completing the magnetic circuit. As the rest of the tape was in contact with the bare feeder, the current would pass along it to the studs; and, therefore, only these under the action of the magnet and in contact with the collecting brushes would become energised. The objections to this system are fairly obvious, consisting of insufficient contact between the tape, studs, and feeder, and impossibility of working at high speeds. In addition, the action of the tape would necessarily be erratic, and anything but reliable.

This is very ancient experimental history, and we come now to a comparatively modern magnetic system which has been to some extent in practical use—viz., the Diatto. Patented in 1894, this system has been tried in both Italy

and France ; judging from reports, with very mixed **results**
Fig. 41 is a diagram of the arrangement. Between **the**
rails are placed the contact studs S, which consist **of a**
circular piece of non-magnetic metal fitted with a **soft iron**
core. To the former is attached a casting (not **shown)**
which very neatly encloses the whole of the working **parts**

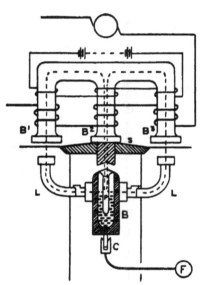

—that is to say, the iron
plunger P and the **mercury**
bath B in which it is placed.
The conical head of the **plun-**
ger is formed of hard, homo-
geneous graphite, and the iron
core just above it is also **fitted**
with the like material. **The**
mercury in the bath—**which is**
made of ebonite—is electrically connected to the **feeder F**
through a clip C, **and the**
whole apparatus is fixed **in a**
small pit, resting, as shown, on

FIG. 41. — DIAGRAM OF DIATTO SURFACE CONTACT SYSTEM.

two projecting iron lugs L L **at**
the sides. These lugs, **curving**
upward through the road bed toward the surface, **form**
a very important part of the apparatus, as will be **shown**
later. The collecting and magnetic arrangements **carried**
on the car, as in some other systems, combine **the**
two functions in the one apparatus as follows. **Under-**
neath the car, running very close to the road surface, **are**
three iron bars B 1, B 2, B 3, in length rather more **than**
equal to the distance between two studs. The centre **one**
makes contact with the latter, acting as the collecting **skate.**

Along the upper surface of the bars, and bolted to them, are a series of electro-magnets, connected in such wise as to make the centre limb a north pole and the two outers south.   The action is as follows : When the front part of the skate reaches a contact stud, the magnetic field passes through the apparatus in the path shown by the dotted lines, and the natural consequence of this is the attraction of the plunger P to the iron core in the stud.   The latter is now energised through the mercury bath, and B 2 picks up the current.   As the car leaves a stop the plunger falls and breaks the circuit.   It will be noted that the magnet has two windings, one in series with the motor, and the other fed from a small battery.   This latter consists of fifteen cells, and is used for exciting the magnets when starting.   Obviously, when the car is standing there is no current passing through the motor, and the magnet is therefore out of action.

The best point about this system is the convenience with which a faulty stud can be replaced.   After disengaging from the surface of the road, all that is necessary is to lift the entire apparatus out.   As it only rests on the lugs L L and comes away easily from the clip C, it is a very simple matter that can be effected in a few minutes.   The magnetic circuit as formed by the above lugs is also a good feature, as it reduces the air gap to a minimum, and therefore causes the action of the plunger to be very decisive.   The misfortunes that have attended the use of this system in Paris appear to have been chiefly due to minor causes, which should easily be rectified.   For instance, in order to give stability to the studs, they were fixed to iron girders attached to the track, with the consequence that the insulation of the system was very much reduced, and leakage in bad weather became considerable. Also, the draining of the pits and the insulation of the feeder cables appear to have been bad, with the result that leakage and shock risks were much accentuated. As regards the actual working of the switch there does not seem to have been much to complain of.

A comparatively recent attempt to solve the surface con-

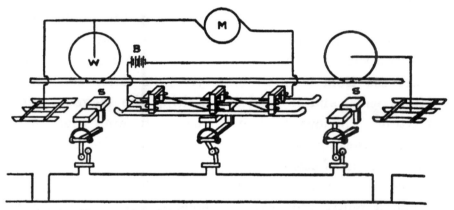

FIG. 42.—DIAGRAM OF DOLTER SURFACE CONTACT SYSTEM.

tact problem is that of the Dolter Electric Traction Company. Their system is magnetic, and is one of the few that may be classed as practicable. It is illustrated in diagram in Fig. 42, from which it will be seen that in certain parts it is akin to the Diatto. Instead of three, there are two skates, which act both as collectors and as pole pieces for the electro-magnets attached to their upper side. When these skates arrive over the iron road contacts S, their magnetic field causes the hinged arm to be attracted upwards, which closes the switch in the manner shown at the centre stud in Fig. 42. The current then passes to the studs, and is picked up by the skates. These are connected together, as shown, at the left-hand end, and the main current flows through the windings of all the magnets in series to the motor M, and to the track through the wheels W. As the skate leaves the studs the switch opens by gravity, the magnetic field that held it up being removed. The installation of the four-cell battery B on the car constitutes an improvement over previous methods. As will be seen from the diagram, it is permanently connected across the magnet windings, with the result that the magnetic field is constant instead of variable according to the load on the

motor. If the load be very light the battery supplies most of the exciting current; but as the power passing to the motor through the series coils increases, the battery does less and less work, until,

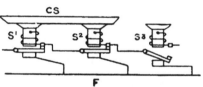

FIG. 43.—DIAGRAM OF WYNNE SUR-
FACE CONTACT SYSTEM.

at a point approximating to full load of the motor, part of the main current is shunted from the coils through the battery, charging it and thus keeping it always in full working order. By this means the strength of the magnetic field is kept approximately constant despite varying load, which removes the objection applicable to many other systems, that the action of the switches is at one moment too decisive, and at another too weak. The details of the switch are, of course, very different to those shown in Fig. 42, which is purely diagrammatic. Contact is made through a carbon block, and the whole arrangement of hinged lever and switch is neatly enclosed in a removable case. The two small leading and trailing skates are no novelty, being employed in most systems. They are connected to earth through the wheels, and their function is to preclude the possibility of a stud being left alive after the car has passed by earthing it and blowing a fuse placed between the stud and the feeder. It is proposed to instal the Dolter system at Torquay and several other towns in Great Britain. The results of its working will be watched with some interest, as it appears one of the best methods yet devised.

Turning now to electrical systems, one of the earliest was that due to Wynne, and, although never in practical use, it is described, as it shows in a simple manner the fundamental principles of all such systems. Fig. 43 illustrates Wynne's suggestion in diagram. C S is the collecting skate carried on the car, S 1, S 2, and S 3 are the road studs, and F is the feeder cable. The iron studs were placed in the centre of the track at equal intervals, and were made magnetic when the car was over them by the coils wound round their lower half and electrically connected to them at the point marked with a dot. Beneath each stud, in a watertight chamber, was placed a hinged keeper provided with a contact piece at its extremity, which engaged with another contact in circuit with the winding. Below this again was a steel spring carrying a disc of soft iron at one end and connected to the feeder cable at the other. In the diagram the car is supposed to be progressing from right to left, and it will be noted that the skate is receiving current from two studs, S 1 and S 2. This constitutes the main principle of the system. For the moment assume that the skate has not yet reached S 1. In this position the switch of that stud will be open. But directly contact is established between S 1 and the collector, the current flowing through the steel spring, iron disc, and switch of S 2 will divide at the latter point, part circulating through the coils of S 2, and the remainder through the connecting wire and coils of S 1. This stud now becomes magnetic, and attracts its keeper upward until it is in the position shown, where it receives current direct from the feeder. As the car moves forward, contact between the skate and S 2 is broken, with the result that no current flows through its winding, and the switch opens by gravity as in S 3. The idea is ingenious and simple, but it is open to objections in practice of a serious nature. Firstly, the car can obviously move only in one direction; if the connections be followed out, it will be seen that a backward movement does not operate the switches. In a contact made between a road stud and a skate it is not always possible to secure good connection; and this might result in the current shunted round the leading stud being too small to exert the necessary magnetic pull on the keeper. In this case the circuit would be broken at the trailing stud, and it is difficult to see how it could be again established without considerable complication. The third and greatest disadvantage is one common to the majority of electrical surface contact systems. There is always the possibility that leakage from the stud to earth may be sufficiently great to keep the iron core magnetised, in which case it would remain alive after the car has passed and constitute a grave source of danger to the public. The latter disadvantage alone was sufficient to kill the system, and although attempts were made, it was found impossible to guarantee absolute immunity from this risk.

Fig. 44 is a diagram of the Johnson-Lundell **surface** contact system, which is of modern origin and **contains** some good features. There is the usual skate **and row** of studs connected through magnetic switches to the **feeder** F, and, in addition, there is a sectional rail S R laid **very** close to, but insulated from one of the track rails, **and two** small contact wheels C C, which bear upon it. A **small** battery is carried on the car for the purpose of energising

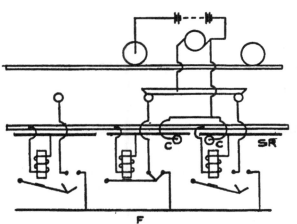

the first **stud** when **starting.** The **diagram** shows that **the** skate picks up the current from **the** road stud, return- ing it viâ the con- tact wheels to the sectional rail, **and** thence through the coil of **the** magnetic **switch**

FIG. 44.—DIAGRAM OF JOHNSON-LUNDELL SURFACE CONTACT SYSTEM.

to earth or the running rails. When the leading contact **wheel** in its forward movement touches a sectional rail the return current is divided, part still going through the first switch and part through the second one, closing the latter **and** making its stud—which shortly comes under the **skate—** alive. With this arrangement it is obvious that the **car** can work in both directions. Its primary object **is to** prevent the possibility of the switches remaining **closed** after the car has passed from the magnetic action **of any** surface leakage through the coils from the stud **to earth.**

It does this most effectually. Only a leakage current passing from the stud to the sectional rail can energise the magnets and keep the switches closed, and this is obviously impossible, as the track rail is placed between the two and must absorb any leakage emanating from the stud. Thus, one of the greatest difficulties of electrically operated surface contact system is surmounted, but at the expense of some convenience in working owing to the presence of the contact wheels and extra rail. There appears to be a certain amount of danger of the track and the sectional rail becoming electrically connected; but as this cuts out the coil, it simply makes the switch incapable of closing. By no possible means can it be left on through electrical action. The insulation of the sectional rail is where trouble may be looked for, as successful working obviously depends entirely upon it. If this can be done thoroughly and permanently so as not to be depreciated by mud and damp there is here, to all appearances, a very good system.

The fundamental idea underlying the mechanical method of surface contact traction is, that while electrical and magnetic apparatus may fail, sometimes from obscure and outwardly trivial causes, mechanical action is tangible, and, therefore, more reliable and certain. Further, a fault in the system is easier rectified; and until something breaks or wears excessively, a well-designed mechanical system cannot fail to act. The trouble experienced with early electrical and magnetic methods gave rise to a large number of attempts to solve the difficulty by mechanical means. The majority are simple, and all are interesting; but space will not permit the mention of more than one or two examples. In 1898 Messrs. Allen and Peard brought out

a novel and, to a certain extent, effective system. An outline
idea of the switch is afforded by Fig. 45, which shows that
it is an essential part of the contact stud itself. At each
box a hole is made through the adjacent rail, and in this is
situated the button B, partly filling up the rail groove at
this point. As will be seen, the button acts upon one end
of the lever L, the other extremity of which is geared to the
contact stud C S. This stud is movable in an upward
direction against the coiled spring S; in its normal position

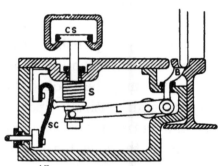

it lies in its bed on a level
with the road surface, being
pressed down by the spring.
The spindle of the contact
stud, which works through
a gland, as shown, is insulated by a bush from the
lever, but it is in electrical
connection with the switch
contact piece S C. A tap-

FIG. 45.—DIAGRAM OF ALLEN AND
PEARD SURFACE CONTACT SYSTEM.

ping from the feeder cable is taken to the laminated copper
bow which forms the other contact of the switch. The action
is very simple, and for this reason it would appear to be
reliable so far as the mechanism goes. The leading wheel of
the car when it reaches the button B presses it down, lifting
the stud against the action of the spring and bringing it
into electrical connection with the feeder through the
contact piece and the laminated bow. Of course, when the
wheel leaves the button the switch tends to open; but here
the skate comes into action, and by its bearing upon the
underneath side of the stud keeps it up and consequently
alive for the entire length of the travel. When the skate

leaves the stud the latter is pulled forcibly down by the spring, and contact is broken with certainty. While possessing the advantage of simplicity, there are a number of practical objections which may be made to this system. The type of skate is unusual and cumbersome, and it would be subjected to considerable wear by the pull of the powerful spring. There is also the certainty of dirt working under the stud, and although enough could not collect to leave the switch in contact, the projection above the road level might be sufficient to cause some obstruction. It is doubtful if

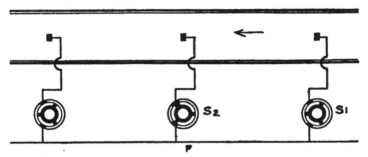

FIG. 46.—DIAGRAM OF KINGSLAND SURFACE CONTACT SYSTEM.

the glands could be made thoroughly water-tight, and failing constant inspection, the box might accumulate sufficient water to make the stud alive when in its lowered position. No doubt the system as illustrated might be much improved upon, but, obviously, radical changes would have to be effected.

Perhaps the most practicable mechanical type of surface contact traction yet devised is the Kingsland, illustrated diagrammatically in Fig. 46. As a whole it is simple and well designed, showing carefully thought-out detail and appreciation of the difficulties inseparable from surface contact traction. It necessitates the use of a shallow conduit,

built up by means of a special rail placed close to one of
the running rails, a gap of five-eighths of an inch being left
between the two heads at the surface.   The road studs are
placed at stated intervals along the track, and opposite each of
these, in a special chamber, is the mechanical switch of which
the contacts are shown in diagram in Fig. 46.   Leaving out
the details, which are rather complicated, these switches
consist of three essential parts—two fixed contacts, a movable one having three projecting arms, and a lever connected
to the latter through a form of ratchet gear.   The contacts
are enclosed in a water-tight box suspended from the roof
of the chamber, and the lever—passing through a gland—
projects upward into the conduit.   The car is provided at
each end of the frame with a striker bar or plough, which
run in the conduit and engage with the switch lever, forcing
it down and operating the switch.   Referring to the diagram,
before the car reaches a stud the switch is in the open
position, its contacts being situated as shown at S 1.   Upon
the plough forcing down the lever of this switch, the movable contact is rotated one-sixth of a revolution, and connection is made between the feeder F and the road stud as
at S 2.   The lever resumes its vertical position by aid of a
spring directly the plough leaves it, but as it works upon a
ratchet principle the switch itself is not affected.   When
the switch is closed the skate touches the stud and collects
current from it.   The progress of the car next brings the
rear striker into action ; the lever is again pressed down,
another one-sixth of a revolution is completed, and the
switch is left open, the lever returning as before to the
vertical position.   The distance apart of the strikers is so
arranged that current is switched on to the leading stud

before the skate reaches it, and is not cut off the trailing one until after the skate has left, thus avoiding sparking at the opening switch. If for any reason the leading switch fails to act, the current is broken at the rear stud, which will not be appreciably harmed by the arc. This arrangement has the additional advantage of drawing attention to the fact that the switch has failed.

The question now arises, which of these three methods is the best for average conditions — electrical, magnetic, or mechanical? What is required is not the best theoretically, but the one which at minimum expense is most suited to the rough usage of every-day work, and at the same time affords the greatest amount of safety to traffic. From evidence in the shape of reports it is clear that the electrical system on the whole cannot compete with either a good magnetic or mechanical one. It will usually be found considerably more expensive both in capital outlay and in maintenance, and more liable to go wrong. Large practical experience has been had with such methods, almost everywhere with the admitted result of failure. Anything approaching safe working requires complication with its attendant evils of heavy outlay and extra attention. The choice must lie between the magnetic and mechanical methods, upon both of which fairly cheap and reliable systems have been based. Experience with mechanical methods is somewhat limited, having so far been confined to experimental work only ; but the possibilities are great if the problem be tackled with thoroughness. The requisite points are simplicity and strength of the working parts, ease of access for renewal, and operation in a manner which interferes as little as possible with the road surface. If at the same time unusual structural difficulties at junctions, etc., can be avoided, and the cost can be approximated to that of the trolley, mechanically switched surface contacts should be a feasible proposition. The field is comparatively new, and it is within the bounds of possibility that, given a fair show, a mechanically switched system might be evolved that would help to pull surface contact traction out of its present discredited position. On the whole, magnetic methods seem to be the most favoured at the moment. They possess an advantage over mechanical switching in having been commercially tried to some extent. That they have not been uniformly successful is well known; but where they have failed it has generally been due to bad management and lack of care in the details, not to any fundamental fault in the method. This is evidently recognised by inventors, as of late years almost all fresh schemes have been based upon magnetic operation of the switches.

CHAPTER XVIII.

CAR BUILDING AND EQUIPMENT.

Trucks—Rigid and Bogie Trucks—The Maximum Traction Truck—
Car Bodies—The Trolley—Tramway Motors—Construction of
Motors—Motor Suspension and Gearing—The Series-Parallel
Controller—Electrical Braking—Construction of the Controller—
The Hand Brake—Power Brakes—Track Brakes—The Westing-
house Magnetic Track Brake—Minor Auxiliaries—Regenerative
Control.

In every respect the modern electric tramcar differs
widely from its horse-drawn prototype. The conditions are
of course quite dissimilar, and in the construction of an
electric car it is necessary to provide for numerous exigencies,
for the most part totally unknown in the days of horse
traction. In no section of the equipment is this more
important than in the truck, the carriage upon which the
motors and the car body are mounted. It is the part which
deals with most of the strains of traction, and upon its
good design and efficient upkeep depend in a great measure
smooth running and low maintenance costs. For many
years America led the way in truck building, and equipped
practically every tramcar to this extent ; however, British
enterprise has at last been brought to bear on the subject,
with the result that there is at present no need to go
abroad for trucks of any description.

The conditions to be met differ considerably, and the
type of truck employed for tramcar work will depend upon

FIG. 47. TRAMCAR BOGIE TRUCK FITTED WITH MAGNET C TRACK BRAKE.

variable quantities, such as the class of car to be used, the gradients and curves of the line, etc. It is impossible to go into the details of each, or to trace the development of truck building from the earliest crude efforts. It is only necessary to note that at the present day trucks may primarily be divided into two classes—those of the rigid four wheel type, and bogie trucks employed for longer cars. The former class is perhaps the most favoured, for the reason that it adapts itself very well to the conditions which obtain in most British tramway systems. The short double decked car is very popular in this country, and for this type there is no advantage in adopting the bogie principle. On the other hand, for long cars such as are employed by the London United Tramways Company the bogie or swivelling truck is a necessity, as, for one reason only, a rigid truck long enough to accommodate the lengthy body could only negotiate curves of the largest radius. The chief disadvantage in employing bogie trucks with the usual two motor equipment lies in the fact that the weight of the car, instead of coming solely upon the driving wheels, as in the four wheel type, is distributed over eight wheels, of which only four are drivers. With an equal weight of car, what is termed the equal traction bogie type will therefore not possess the driving force of the rigid truck, as only half the weight bears upon the driving wheels. This difficulty was soon discovered, and to remedy it the maximum traction bogie truck was invented, which by a different design allows about 80 per cent. of the weight to be borne by the driving wheels, the remaining 20 per cent. being taken by the pony wheels. As the entire weight of the car is useful for traction purposes, the four wheel rigid truck is the better for hill climbing or for greasy track.

The bogie truck can never equal it in this respect **unless it** be provided with four instead of two motors.

The essentials of a good truck are that it should **combine** lightness with strength and rigidity, that it should **give** easy access to all parts, that its springs should **be so** proportioned as to ensure easy running at all **loads,** and that its bearing parts should be self-lubricating, and call for little attention. In addition, there are **more** detailed considerations which must be taken into **account** in its design, such as its proper behaviour on curves, **the** manner in which braking gear can be applied, **and the** means provided for the attachment of the car body.

To the most casual observer it will be clear that **great** alterations have been made in car body design with **the** introduction of electric traction. Besides enhanced com- fort and a more elaborate decorative scheme, the bodies **are** much stronger in every part, in order to meet the **strains** set up by higher speeds and heavier loads. Where **four** wheel rigid trucks are used the car body is securely **bolted** to the truck ; it is, however, easily removable, this **being a** particularly important feature. With bogie trucks **the** body rests upon what are termed "rub plates," **so as to** allow of swivelling action when rounding curves. The **pull** of the truck is taken by suitable pins, which are **designed** solely for this purpose, and not to secure the body **to the** truck, as in the case of the four wheeled variety. **Trap** doors are provided in the floor of the body over each **motor,** to allow of cleaning and easy access for small **repairs.** Where the overhead trolley system is used **the roofs are** made specially strong in order to cope with **the strains set** up by the trolley. These are often considerable, **particularly** when the car lurches, and as the leverage of the trolley arm is very great, there is a heavy strain on its base, which must be provided for by thoroughly sound fixing. It is

FIG. 48.—TRAMWAY MOTOR, OPEN, SHOWING ARMATURE AND FIELD MAGNETS.

customary, and necessary in the case of top seat cars, for the trolley arm to be attached by means of a swivelling arrangement to the top of an iron standard fixed on the

o 2

roof. This pillar contains a strong helical spring, which tends to press the arm against the wire with a pressure of from 15 to 25 lbs. according to the system of suspension. The cable conveying current to the motors also passes down the standard, which is connected to earth through an indicating device so as to safeguard passengers against shock if, from wear of the insulation, the live cable should come into contact with it. The trolley head is that part which makes connection with the overhead wire, and it consists of a grooved gun-metal wheel running in bearings designed for attachment to the pole. In course of time it becomes necessary to renew the wheel, which wears in the centre of the groove, and provision is made to render it easily detachable.

The standard equipment for tramway work consists of two series wound 25 to 40 h.-p. motors, specially designed to meet the extremely onerous conditions which prevail. Fig. 48 shows a well-known make of tramway motor with its field magnet in the open position. It will be noticed that the casing which forms part of the magnetic circuit is hinged, so that when it is necessary to get at the armature or connections ; all that is required is to open the frame and lay the whole interior bare. The casing is of cast steel, and according to the make the poles either form an integral part of it or are built up of steel laminations and fastened to the yoke by bolts. The field coils are wound on formers, and being simply slipped over the poles, are readily renewable. The armature is always of the slotted type, that is to say, the coils are bedded in slots cut radially towards the centre of the core. Like the field coils, they are wound on formers, and after being slipped into the slots are retained in position by steel wire bands wound circumferentially

round the armature in shallow grooves.   Insulation is very
carefully attended to, varnished cloth, tape, fullerboard,
mica, and asbestos being variously used.   The commutator
is always of large diameter and fairly short, in order not to
increase the space required unduly.   The segments are of
hard drawn copper, with a good wearing depth so as to
allow of trueing up, and the insulation between them is, as
usual, mica.   Carbon brushes are invariably employed, and
set on the commutator radially and without any rake, so
that their duty is performed equally well whichever
direction the armature is rotating in.   The brush holders
are of a simple nature, and are designed in conjunction
with the case, so that renewal is easily effected.   Wherever
necessary, such as over the commutator, hand holes are
provided, the covers being attached by bolts and a tight
waterproof fit ensured by a thick felt insertion.   The
armature bearings are of course self-lubricating, either oil,
grease, or a mixture of both being employed.   The larger
bearings, shown in the illustration, are for the accommodation of the car axles, which in the form of attachment
generally considered preferable and known as " nose suspension," take a great part of the weight of the motors.   In
this form of suspension the side of the motor remote from
the axle is attached to the truck by a cross bar and springs,
and it is claimed for it that the motor is more easily removed
for repair and that the spur gears work more satisfactorily
than with any other type of fixing.   The alternative methods
consist of supporting the motor at its ends by the aid of two
horizontal bars or brackets attached to the frame, or by a
bracket placed centrally underneath it.   The motion of the
armature is transmitted to the axle by steel gears, one, the

pinion, being keyed to the motor shaft, and the other to the axle. The gear is run in grease; and the teeth of the spur wheels are large. The usual width of face is five inches,

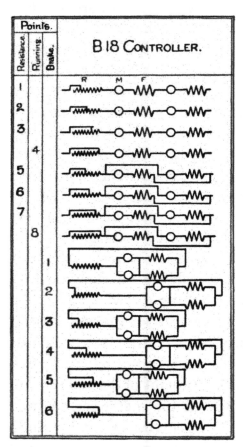

but a variation of half an inch either way is sometimes made. The ratio of speed reduction is usually from 1 to 4, to 1 to 5; but this will differ according to circumstances.

The essential requirements of a tramway motor are that it should be impervious to dirt and damp, as far as possible accessible in every part, easily removable, and thoroughly sound in construction. In its operation it must be capable of exerting a large starting torque, must stand temporary overload without undue heating, and must run sparklessly at all loads and in either direction. The evolution of a motor capable of fulfilling

FIG. 49.—DIAGRAM OF CONNECTIONS OF BRITISH THOMSON-HOUSTON TRAM-WAY CONTROLLER.

these requirements has been a difficult process, but it has been achieved in conjunction with a high efficiency in converting electrical into mechanical power.

However good the motor, direct current electric traction

could never have attained its present efficiency without the system of series-parallel control made possible by the modern controller. It has been pointed out that at starting a resistance must be inserted in series with a motor, and the same course, with its attendant inefficiency, would be necessary to reduce the speed. The modern method of controlling the two motors which form the usual tramway equipment is shown in diagram, Fig. 49. Assuming the use of the British Thomson-Houston Co.'s B 18 Type controller, the connections of the motors M, their series windings F, and the series resistance R at the various points are as follows: There are in all fourteen combinations, eight of which are used in ordinary running and six when braking the car. In starting, the handle is moved to position No. 1, and the two motors and their field windings are connected in series with each other and with the greater part of the resistance. As each motor is wound for the full line pressure—say 500 volts—they start slowly, something less than half their proper voltage being applied to them, and the balance absorbed in the resistance. Notches 1, 2 and 3 are passed over in a few seconds, the effect being simply to cut out the resistance in sections, as shown. These notches are termed resistance points, and the controller must not be left at them for any length of time, as the resistance would heat and give rise to waste. At position 4 all the resistance is cut out, the motors still running in series, and, therefore, at a low speed but fairly efficiently. As there is no resistance to heat up, and but little waste, this point may be used for any length of time when slow speed is required, and it is therefore termed a running point. The next step, point 5, is an important one, as here the change from series to parallel is effected.   Half the resistance is put into circuit, and at the same time the armatures and fields of the two motors are thrown each direct across the line pressure, receiving the full 500 volts less that amount absorbed in the resistance.   At points 6 and 7 this is reduced in steps, and at point 8 it is entirely cut out, the motors running with 500 volts at their terminals, and consequently at full speed.   This is the normal running point; the two preceding ones are resistance steps as before, and intended only for passing over quickly.   In the second half of the control system one of the most valuable characteristics of the electric motor—its reversibility—is made use of.   When the car is running by its own momentum preparatory to stopping or down hill the motor armatures are mechanically rotated, and if the + and — brushes be connected together so as to form a circuit, the machines will begin to generate current and thus retard the motion of the car.   It would be unwise to couple the brushes with no resistance in circuit when running at a high speed, as the current that would flow would be excessive, and, besides exercising a too violent braking effort, it would injure the armatures by overheating and the commutators and brushes by sparking.   At the ninth or first braking notch the terminals of the two motors—still in parallel—are therefore connected together through the entire resistance, which allows just sufficient current to flow to exercise a small retarding effort.   At each of the remaining positions the resistance is reduced in steps, until at the sixth braking notch the motors are directly short circuited, the whole of the resistance being cut out.   Returning now to the first half of the controller, the part which is used for accelerating or running at reduced speed. Obviously, the connection of the motors in series is much

FIG. 50.—WESTINGHOUSE TRAMWAY CONTROLLER WITH CASE AND SPARK GUARDS OPEN.

more economical than the use of resistance. In the latter case the power absorbed is dissipated in valueless heat, but in the former it is converted into mechanical work, the only waste occurring being occasioned by the series resistance. This is small compared with the inefficiency of a variable resistance in series with the motors permanently connected in parallel—more particularly as the resistance is in use only for a very short time—and, therefore, series-parallel control affords economic advantages where direct current working is concerned which are impossible to attain by any other practicable means. The controller, of which the connections are shown in Fig. 49, is a simple one, and for this reason it is given as an example. It affords a good idea of main principles, perhaps better than would a more complicated type. From this point of view there is therefore nothing to be gained by considering other and less simple arrangements; but it should be borne in mind that many tramway controllers are of a more complicated nature, providing for a greater range of efficient speed control and possessing additional braking points. Where four motors are used on the car the connections are, of course, different; and the employment of a magnetic brake will also further complicate matters.

It is necessary now to examine the controller itself. One built by the British Westinghouse Co. is illustrated in Fig. 50 with the case open to show the working parts. Although at first sight complicated, it will be seen that it consists of two fundamental parts: a central rod carrying a series of segmental contacts, and on the left inside the case a number of " fingers," as they are termed, with which the segments make connection. The movement of the handle rotates the " drum," and by aid of the varying lengths and positions of the segments, makes the connections in their proper sequence. On the right is seen the reversing lever, with its contacts beneath it. According to the position of this the car will run either forward or backward, the operation of the main lever producing the same conditions of running in either direction. This particular controller possesses five steps for the series and three for the parallel connection of the motors, and in the braking position there are seven steps. The spark guards are shown open on the extreme left. They are so arranged that when closed they surround the contact fingers and render arcing from one to the other or to any part of the frame impossible. As a certain amount of sparking occurs at the fingers when breaking circuit, these are provided with special renewable tips. In other controllers, notably those of the British Thomson-Houston Co., the magnetic blow out principle is utilised, and by the aid of a number of small coils, or one large one carrying the main current, the arc is broken directly it is formed.

The increasing number of tramway accidents due to faulty or ineffectual brakes has drawn public attention to the subject, and made patent the fact—long appreciated by engineers—that methods of stopping the car are of equal importance to methods of propulsion. The weight of an electric car when fully loaded is very considerable, and this, in conjunction with the increased speed, calls for much more powerful arrangements than were found necessary with the horse-drawn car. The hand brake with which nearly every car is equipped is a legacy of horse traction, but in the electric tramway it, of course, assumes a more powerful shape. It is operated from the driver's platform by the agency of chains and levers, the brake shoes being pressed against the tread of the wheels, thus creating friction which retards the car. With all brakes of this type it is possible to skid the wheels—that is to say, to lock them so that the car slides along the rails—but this is always extremely undesirable. With any class of wheel brake the best effect is produced when the wheels are slowed down to a point just short of where locking and skidding takes place. As recent accidents have proved, a car with locked wheels will slide a considerable distance before pulling up, particularly if the track is greasy, and, in addition, skidding produces flats on the wheels which militate against good braking and create noise and pounding. One of the elementary duties of a motor man is to learn how to stop his car without skidding the wheels by a too sudden or powerful application of the brake.

Besides the hand brake, it is necessary in almost all cases to provide additional means of retardation, which may take the form of a power-operated wheel brake or one of the many types of track brakes. On the Continent an air brake, such as the Christensen or Westinghouse, is favoured, and the results are as a rule satisfactory. On the other hand, the cost of such brakes is excessive, the equipment comprising an air pump and reservoir in addition to the numerous pipes and valves. In view of this, the advantages of the air brake are more apparent in tramway systems employing trailer cars than in the usual British system of one car, and hence it is not by any means common in this country. The track brake is a much cheaper contrivance, and a large number of different types have been designed. The aim of all these is to set up rubbing friction between the car and the rail, thus securing a retarding effort independent of braking action on the wheels. It is usually obtained by the aid of shoes, which are attached flexibly to the truck and pressed down on to the rails by powerful levers; but pneumatically-operated track or slipper brakes have been employed. While favourable results have been attained by many such brakes they suffer from one initial disadvantage in that their operation tends to take the weight of the car off the wheels. When the wheel brakes are worked in conjunction this is highly undesirable, as the pressure being to some extent removed, the wheels are more liable to skid when the

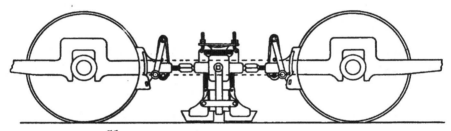

FIG. 51.—WESTINGHOUSE MAGNETIC TRACK BRAKE.

shoes are applied. Further, cases have been known where a forcible application of the slipper brake has derailed the car through taking the weight off the wheels. These objections are entirely removed in the Westinghouse magnetic track brake, a description of which is given, as it is without doubt the best slipper brake yet produced. It is illustrated in Fig. 51 attached to a single truck, and it will be seen that it possesses three essential parts, namely, a track magnet, brake blocks acting on the wheels, and a link mechanism connecting the two together. The brake is suspended flexibly from the truck by means of steel springs, and the equipment consists of two brakes—one on each side—for a single truck, and four for a bogie. The soft steel shoes of the track magnet run about $\frac{1}{4}$ inch clear of the rail, and are detachable to facilitate renewal. They act as poles to the electro-magnet hung just above them parallel to the track. The magnet is energised by the motors acting as generators, the circuit being completed through the coil when the controller is on the braking notches. The excitation of the magnet results in adhesion of the shoes to the rails and a strong downward pull, which is transmitted through the gear to the wheel brake shoes. Instead of taking the weight of the car off the wheels its adhesion to the track is increased by the magnetic pull; and, in addition to the braking effort of the track and wheel shoes, there is the usual retardation due to the motors acting as generators. It will be seen that with this brake the forces which come into play when the controller is moved to the braking notches are both numerous and powerful. The result is a retarding effect on the car sufficient to pull it up in a short distance under any conditions of speed or grade. The higher the speed, the more powerful is the braking, as the E.M.F. generated by the motors is proportional to the speed. At the same time, the brake will effectively retard the car at the slowest speed, releasing directly it stops, but resuming its action the moment it begins to move again. The Westinghouse brake need not necessarily be connected to the controller brake notches, but such is the general practice as opposed to installing a separate switch. While effective as an emergency stop it is also employed in ordinary working, its capability of stopping the car at slow speeds making this possible. Magnetic track brakes dependent for their action on a supply of current taken from the line have been devised, but they are now obsolete, having been found

wanting in practice.   They are obviously unsafe, since, if
the trolley left the wire or the current failed while the car is
on a down grade, the brake would become totally unoperative.

There are numerous minor auxiliaries employed on a
car, but in a cursory review of the subject the majority of
these do not warrant attention.   Among the most important is the circuit breaker, generally placed on the roof
over the driver's head.   The motor current passes through
this switch, which automatically opens in the usual way
should the current become excessive.   Sanding gear is
another important adjunct.   By pressure of the foot on a
button the driver can release a quantity of sand, which falls
on the rails in front of the wheels, preventing skidding and
giving them a better grip on the track in greasy weather.
This is sometimes necessary at starting, or when the brakes
are sharply applied.

The lighting arrangements are controlled by switches
grouped together in a convenient position, and 100-volt
lamps are usually employed, connected five in series.

The regenerative system of control, to which attention
was drawn in a preceding chapter, employs shunt instead
of series wound motors, and by this means the car is enabled
to return energy to the line when running down hill or
stopping.   A series motor obviously cannot do this, since
when required to run as a generator it will not excite its
fields unless the armature circuit be completed through a
fairly low resistance.   Where it is desired to return power
to the line by means of the motors acting as generators
they must not be connected to the line until they are
generating a pressure at least equal to that which normally
obtains.   With the shunt motor this is easily arranged,

requiring only adjustment of the current in the shunt circuit. The chief advantage of the regenerative system of control lies in the fact that braking is performed by the generation of electric power, which is returned to the line, instead of by wasteful friction. The saving in power consumption may by this means amount to as much as from 15 to 20 per cent., and, in addition, the retardation, while forcible, is free from objectionable jerks. The range of efficient speed control is much wider than that possible with series wound motors, and the acceleration is smooth and regular. A mechanical brake is, of course, employed to bring the car to a dead stop; but the motors themselves perform the braking down to a very low speed. It will be clear that in the event of the trolley leaving the overhead wire the braking system, which is dependent upon the return of current to the line, will become useless. In the best known regenerative system—viz., that invented by Mr. J. Raworth—this difficulty is overcome by a very simple expedient. A series coil is added to the magnet windings, and when the shunt current fails owing to the trolley leaving the wire, this coil is automatically switched in, together with resistances through which the motors are short-circuited. The braking conditions of an ordinary series motor equipment are thus faithfully reproduced, with results that are in every way as efficient. This system of control is an extremely interesting one, and its details will repay thorough investigation. Being new, it has so far found a place in few text books; but many excellent articles have from time to time appeared in the technical press, and from these an intimate knowledge of the theory of the system and its commercial value may be readily obtained.

CHAPTER XIX.

ELECTRIC RAILWAYS.

Railway and Tramway Practice—The Electrical Locomotive—Main and Suburban Lines—Direct Current Third-Rail System—Converted Alternating Current System—Three-Phase System—Single-Phase System—Multiple Unit System of Control.

UP to a certain point electric railway engineering has developed along the same lines as tramway practice. Improvements in trucks, car bodies, motors, and equipments generally have had an equal effect in developing both, and for several years the methods of the one were, with but slight variations, the methods of the other. So long as powers remained practically equal—that is to say, so long as small and light electric railways only were attempted—tramway methods and equipments sufficed. But, with the advent of the heavy and high speed electric train, calling for a large amount of power often to be distributed over a considerable distance, tramway and railway practice began to diverge, until at the present day the latter, striking out along its own line, employs methods very different in detail to the former. Following the precedent of the steam railway, motor-driven locomotives were used for hauling trains of cars, and at an early date it was found that, as compared with steam, greater average speed and quicker acceleration from standing to full running speed could be efficiently obtained. In addition, the truth of the theory became apparent that it was more economical to generate power by the aid of highly efficient boilers and engines in a central power station, transmitting it to the train electrically, than to evolve it in a comparatively inefficient small unit, such as the steam locomotive, under conditions of economy and convenience which are not to be compared. Although at first sight it may appear that the above somewhat obvious advantages should apply equally to all classes of railways, this is by no means the case. In the present state of knowledge we may confidently predict success for almost any short suburban or interurban electrical line carrying a frequent service. But when we approach the question of main line electrification the problem assumes a totally different aspect, and one, at present, less favourable to electricity. The difficulties are chiefly financial, and involve very complicated questions which cannot be entered into here. It will be sufficient to point out that a frequent and short-distance service with a comparatively insignificant length of transmission line, makes much better use of the capital expended than a main line of considerable length over which heavy and infrequent trains are run. In the latter case less use is made of the feeders and converting gear, also, in all probability, of the generating plant : and further, the losses in power transmission and conversion should be proportionately greater.

Leaving the broad question of economics, we may now turn to the methods employed in actual practice. Of these there are four, each particularly suitable for a certain set of conditions. Where the line is very short, and the power required comparatively small, it is usual to employ a direct current 500-volt third-rail system. Power is generated at this pressure and transmitted to the track feeding points direct, without transformation of any sort. Two examples of this method are provided by the Great Northern and City Railway and the Waterloo and City line. Primarily, everything is as simple as it can be, complications only arising when it becomes necessary to use negative boosters and battery sub-stations for the purposes respectively of reducing the pressure drop in the return circuit, and to cope with rapidly varying loads at parts of the line remote from the generating station. The system is comparable to a small tramway undertaking in which direct current is supplied straight from the generators. The second method is the most usual, being that upon which the majority of the London tube lines and the District and Metropolitan Railways are operated. These lines are dealt with in the succeeding chapters, and it is only necessary here to outline the general principles. On account of the distance over which power has to be transmitted, polyphase alternating current is generated at pressures ranging from 5,000 to 11,000 volts. In this form the power reaches the sub-stations, which are situated at various suitable points along the line. Here it is first transformed down, and then converted by the aid of rotary machines to direct current at approximately 500 volts. The third rail is fed at this pressure, and, therefore, the rolling stock equipments will be similar in all essential details to those employed in the system first described.

In the two remaining systems alternating current only is used, and both offer advantages over direct current for main line or long distance work. It has now been proved beyond doubt that direct current at the usual pressure of

500 volts is totally unsuited for heavy main line railway service. At this pressure the loss in distribution would be very heavy unless everywhere the conductors were of unusually ample capacity, and here the question of cost and interest on capital comes in. The obvious remedy of raising the pressure to several thousand volts cannot be entertained, since direct current motors to work at this pressure are unobtainable; and, therefore, it becomes necessary to fall back upon an exclusively alternating current system with its economy of high pressure transmission and ease of transformation to pressures suitable for application to the motors. In the first alternating current system, polyphase currents are generated at pressures ranging from 3,000 to 20,000 volts, and in this form power is transmitted to the overhead wires from which the train takes its supply. A third rail at such pressures is of course out of the question, and overhead wires supported in a special manner have to be used. On the train, alternating current transformers are carried, and these convert the high pressure three phase current to low pressure which is taken direct to three phase motors. A variant of the system is to employ transforming stations which feed the trolley wires with low pressure current, thus obviating the necessity of carrying transformers on the train and rendering the system much more safe. Polyphase currents for railways are in favour on the Continent, particularly in Switzerland; but as yet there is no instance of their use in Great Britain, and in view of recent developments it does not appear likely that there ever will be. Among the chief polyphase lines are those at Valtellina and between Berlin and Zossen. The system was recommended for the

Metropolitan Railway of London, but as will be remembered, arbitration proceedings between the District and Metropolitan Companies resulted in abandonment of the idea in order to give the two lines the necessary uniformity in their methods.

The second alternating current system employs a single-phase current which is supplied to the overhead wire at high pressure and transformed down on the train. It is the latest development of electric traction, and is only rendered possible by the improvements which have recently been made in the special class of motor employed. Fuller particulars of this will be found in a succeeding chapter. Several suburban lines of the London Brighton and South Coast Railway are being converted to electrical working on this system, and there are a large number of instances in America and a few in Italy where the single-phase electrical railway has proved a thorough commercial and engineering success. To all appearances this is the system of the future, and the next decade should see its use very widely extended.

One of the most important phases of development in electric railway practice was the introduction of the Multiple Unit system by Sprague some ten years ago. It abolishes the locomotive, and equips two or more cars of the train with motors simultaneously controlled from one point. The results are more favourable in every way: the equipment is on the whole cheaper and of less weight, and better acceleration is obtained combined with a much more flexible system. The chief consideration in high speed train operation is the proportion of the weight borne by the driving wheels, and consequently their adhesion to the track.   To obtain the fastest possible train movement between stations it is necessary to increase the proportion of the weight on the driving wheels to a maximum ; and the ease with which this may be effected by the employment of the Multiple Unit system constitutes one of the chief advantages of electric railway traction.   For high acceleration at starting, it is necessary when a locomotive is employed that it should be of considerable weight in order to give the requisite adhesion to the rails.   Much of this weight is unnecessary for other purposes, and besides increasing the cost of the locomotive, it absorbs a certain amount of power in its propulsion.   In the Multiple Unit system the extra dead weight is done away with, the capital outlay is less, and the running cost is lower.   Turning to details of operation, the balance is still in favour of the more modern system.   There is less hammering at the rail joints, owing to the weight being better distributed, and this materially reduces the maintenance cost of every part of the equipment.   The train is readily split up into self propelling sections, affording both flexibility and economy in conducting the traffic.   The operations at termini are simplified and cheapened because the train can be controlled from either end.   There are many further advantages of a minor nature that might be cited, but the above should be sufficient to show broadly where the superiority of the Multiple Unit system lies.   It has replaced the locomotive on the Central London Railway, showing better results, and it is wherever possible employed on both the District and Metropolitan Railways.   All the new tube lines are equipped exclusively upon this system, which alone makes possible their frequent services and high schedule speed. A special form of controller is required, depending upon automatic switches of a rather complicated nature. This is dealt with in a later chapter in conjunction with other details of train equipment.

CHAPTER XX.

THE Lot's Road, Chelsea, power house of the Underground
Electric Railways Company of London, Limited, is the
second largest generating station in the world, being
surpassed only by the works of the Interborough Rapid
Transit Company of New York.   On broad lines, the equipment is typical of modern practice in large power station
design, and a description of the plant and methods employed
will therefore cover both the Railway and the Power Supply
Company's works.

Lot's Road at the present time feeds the Great Northern,
Piccadilly and Brompton Railway, the Charing Cross, Euston
and Hampstead Railway, and the Baker Street and Waterloo
Railway.   In addition, it supplies power to the District Railway, including the southern half of the Inner Circle; and it
is destined to feed the Edgware and Hampstead Railway
when it is completed.   The northern half of the Inner Circle,
which is the property of the Metropolitan Railway Company, is supplied from their Neasden generating station,
comparable to that at Chelsea in everything but in size.

The site upon which Lot's Road stands was selected for reasons that always have an important bearing upon the choice. Its proximity to the river Thames allows coal to be brought to the works by barge at a comparatively low cost for carriage, and it also affords ample facilities for obtaining a cheap and reliable supply of water for condensing purposes. Many difficulties were experienced in the erection of such a vast and heavy building on marshy soil (the self-supporting steel framework alone weighs about 6,000 tons), but they were overcome by dint of perseverance, and the power house now stands in a position well worth the trouble and money expended upon it. Being foreign to the purpose in hand of dealing solely with the electrical equipment, it is not proposed to go into details of the steam raising plant and turbines. It will suffice to give the leading particulars of this portion of the installation.

The boiler house is built in two stories, and will ultimately contain eighty water-tube boilers of the well known Babcock and Wilcox type, suitable for working at a steam pressure of 175 lbs. per square inch. They are fitted with electrically driven mechanical stokers, to which coal is delivered from the overhead bunkers by automatic machinery. In the same manner, the ashes from the fires are carried away by special plant installed for the purpose. In a power house of this size it is essential that all such operations should be performed by machinery, the amount of coal and ash handled daily being very considerable.

A simple arrangement of pipes—carried where possible under the floors—conveys steam to the turbines, of which there are at present eight sets installed in the engine room. They are of very large size, generating approximately 7,500 brake horse-power at full load and running at a speed of 1,000 revolutions per minute. They are capable of dealing with a heavy temporary overload ; a most necessary characteristic for the class of work they are employed on. Directly connected to the turbine shaft by aid of a flexible claw coupling is the generator, rated at a full load capacity of 5,500 kilowatts. It is of the rotating field-magnet type, and generates a three phase alternating current of a frequency of $33\frac{1}{3}$ cycles per second at a pressure of 11,000 volts. The speed being high, the field magnets of these machines have only four poles, made from a solid forging of Whitworth fluid pressed steel, of great strength and good magnetic properties. The stationary element or armature is built up in the usual way of slotted laminations, enclosed in a heavy cast iron case provided with ventilating holes and divided horizontally. The winding takes the shape of copper bars very carefully insulated from the core, the high pressure of 11,000 volts calling for extraordinary precautions in this respect. In order to make sure of the soundness of their insulation these machines were subjected to an electrical pressure of 30,000 volts between the coils and the core after completion. Their efficiency as converters of mechanical into electrical energy is good, the guarantee figures being 97·25 per cent. at full load, and 90 per cent. at quarter load. Both the turbines and generators were manufactured by the British Westinghouse Electric and Manufacturing Co., Limited, at their Trafford Park Works, Manchester, the firm who were also responsible for the 3,500 k.w. turbo-generator sets installed at the Neasden power station of the Metropolitan Railway.

The exciting current for the field magnets of the Chelsea generators is considerable owing to their large size. Each field winding requires about 180 amperes at 125 volts, and this is normally supplied by four steam driven direct current sets consisting of Allen vertical, enclosed, high speed engines, coupled to 125 k.w. British Thomson-Houston dynamos. There is also a storage battery which is used for supplying current for operating the main switches, and a 125 k.w. motor-generator taking alternating current on the motor side at 220 volts and generating direct current at 125 volts for charging the battery. The majority of the auxiliaries such as pumps, mechanical stokers, etc., are driven by three-phase induction motors.

The switch galleries occupy the whole of one side of the engine room and are three in number, being placed one above the other. The large output and the high pressure of 11,000 volts necessitate very different arrangements to those found in the type of alternating current station dealt with in Chapter IV. The complication is much greater owing to the fact that the switches are for the most part too large to be operated by hand, and the pressure calls for unusual precaution against breakdown as well as the thorough safeguarding of the employees from shock which must of necessity prove fatal. At Chelsea there is of course a vast amount of switchgear of different classes, but interest is mainly centred in the control boards situated on the middle gallery. Of these there are two, one being allotted to the generators and the other to the feeders. The former consists of a desk upon which are fixed various small switches, and a series of vertical panels containing the measuring instruments. Each generator requires for its control one high tension main switch and several auxiliaries; and the former, and some of the latter, are necessarily very large and incapable of being grouped together in a manner convenient for hand operation. The only course open is the one which has been taken, namely, to place the switches themselves in some suitable position and to actuate them electrically from one common centre by means of a low pressure circuit controlled at the board, and special operating mechanism attached to the switches. As before stated, direct current for this purpose is supplied by a 125-volt battery. The closing of the small switch on the inclined board representing the generator main switch completes the 125-volt circuit through the mechanism of the latter, causing it to close at once. The field regulating resistance switches of the alternators are necessarily massive, and these too are operated at a distance from the control board by the same means. Both these and the main switches will be mentioned later; for the present attention should be confined to the control board itself. On each section of the desk devoted to a generator are two lamps, which indicate respectively whether the main switch is closed or open. This is necessary, as the switches themselves are some distance from the board, out of sight, and it is simply provided for by means of an auxiliary circuit. The switch that controls the alternator field magnet circuit is itself mounted on this board and is operated by hand. There is the usual synchronising gear for indicating the right moment at which to close the main switch when the machines are being paralleled, a signal push to convey telegraphic orders to the turbine attendants, and a switch for controlling the speed of the set. This latter is installed in connection with a small motor geared to the turbine governor, the position of which it controls, thus varying the speed. In the majority of generating stations it is necessary to rely for speed variation on valve or governor adjustments made by the engine-driver according to the orders he receives from the switchboard. By the above arrangement this course is rendered unnecessary, the regulation being entirely in the hands of the switchboard operator himself. It is an obvious improvement which makes both for safety and convenience in working.

The instruments provided for each of the generators include three main ammeters—one in each phase—a main voltmeter scaled to indicate in the region of 11,000 volts, but actually working at a low pressure obtained by a small transformer of which the primary is connected to the generator; a power factor indicator to show the phase relationship between the current and the voltage, that is to say, to indicate the lag of the former, an ammeter in the field magnet circuit, and an integrating wattmeter to record the number of units generated by the machine. This is practically the complete equipment for each generator, and owing to the system of remote control above described it occupies but little space. The feeder control board is situated on the same gallery opposite the generator board. It consists of seventeen panels, each of which contains the apparatus for four feeders. Every feeder is controlled by a main switch of a similar type to those employed with the generators excepting that it is arranged to open automatically in the event of a heavy overload. Signal lamps show whether the switch is open or closed, and in addition a gong is provided to give audible warning when a switch opens its circuit. Each feeder possesses the usual ammeters and voltmeters, and the whole apparatus is mounted on vertical panels instead of partly on a desk as with the generators. An entirely separate switchboard is provided for the control of the auxiliary gear used in the station.

The main oil-switches were—like the rest of the switchgear—supplied by the British Thomson-Houston Co., Ltd. Their construction and method of operation may be outlined as follows. Each phase of the circuit is allotted a brick chamber to itself, and in each cell are two pots filled with oil and containing contacts at the bottom. The switch rods, carrying a bridging piece at their ends, descend vertically into these pots, and at the lowest point of their travel the bridge piece completes the circuit through the two contacts. The whole of this part of the gear is immersed in insulating oil, so that when· the switch is opened with a load on there is no possibility of an arc being set up between the contacts. In addition to suppressing the arc the oil effectively insulates the live parts from earth. The six vertical rods—two to each phase—are mechanically coupled together by an insulated crosshead, which is raised or lowered when the switch is opened or closed by powerful helical springs situated on the top of the brick and stone chamber. The motor which actuates the mechanism is also placed here. The separation of each phase of the circuit is very complete, the idea being that one might be totally destroyed without communicating the damage to the others or in any way affecting their working. There are about 100 of these switches installed at Lot's Road in various parts of the building, those connected with the generators being placed on the bottom gallery, the feeder switches on the top gallery, and others connected in the 'bus bar circuit on the middle gallery. Despite the manner in which they are thus scattered about, no difficulty is experienced in their control from the one central position on the middle gallery. The convenience and cheapness of such an arrangement is very great; if any system but that of remote control were used, the working of the station would be at once a more costly and complicated matter.

In the case of the motor-driven field regulating switches for the main generating sets the motor is geared to the switch arm, and in accordance with the operation of the controller placed on the generator control board it will start, stop, or reverse its direction of running. At each end of the range of contacts a limit stop is placed, and by means of this current is automatically cut off the motor when the switch reaches the end of its travel, thus obviating any damage that might be occasioned by carelessness on the part of the switchboard attendant. These switches are situated on the middle gallery, and there is one for each of the main generators.

The 'bus bar arrangements in a station of the size and high working pressure of Lot's Road are very different to those which obtain in a small and comparatively low pressure alternating current plant. It is highly essential that they should be thoroughly protected in order to minimise the risk of shock, but at the same time they must be get-at-able for the purposes of cleaning the insulators and the carrying out of any alterations that may be necessary. These conditions are now universally met by methods similar to those adopted at Chelsea. Here, the main 'bus bars are placed in separate brick chambers provided with iron inspection doors, being carried on the walls by an efficient type of insulator. They are electrically broken up into five sections, separable from one another by oil switches, and by this arrangement it is possible to work all the generators in parallel on the whole length of 'bus bar or to split up the latter and work them two in parallel on each section. The feeders to the substations are duplicated, and are so arranged with but two exceptions that each cable running to any one substation is connected to a different section of the 'bus bars. By this expedient it is possible to shut down a section of the 'bus bars without discontinuing the supply to the substations fed from it, and it is further possible to split the whole or part of the system into independent sections supplied from separate generators. The advantage of these arrangements may not at first sight be apparent, but in electricity supply of any description many remote contingencies of mishap have to be provided for, and it is always necessary that a generating and distributing system should be so designed as to be readily adaptable to extraordinary conditions.

The 11,000 volt feeders running from Chelsea to the various substations are of the three core (one for each phase) paper insulated, lead covered type. They were subjected to a pressure test of 33,000 volts at the factory, and after installation and jointing they were again tested to 22,000 volts. To travellers on the District the method of fixing to the walls of tunnels and cuttings will be well known. The cables are placed one above the other in iron troughs fixed to the walls by aid of brackets. These troughs are so arranged that they will turn down, and expose the cable in any part should it have to be removed or repaired.

When all the lines it is to operate are completed Lot's Road will supply power to 24 substations of various sizes. These will have a total capacity of 87,500 k.w. and an average transforming and converting capacity of 3,646 k.w. each. It is not necessary here to enter into details of their equipment, as this is substantially the same as that of the Metropolitan Railway substations which will be dealt with in the next chapter. For the same reason the track and rolling stock need not be touched upon excepting to mention the following distinctive features of the District car equipment. With all trains of the ordinary seven car length three motor coaches are used ; one at each end and one in the centre. Such a train unloaded weighs 137½ tons, and will seat 364 passengers. Each motor car is equipped with two 200 h.-p. series wound British Thomson-Houston railway motors, and there is therefore 1,200 h.-p. available for propelling a seven coach train. The power at starting may be considerably greater ; the above-mentioned figure is the amount the motors may be safely called upon to develop in continuous running. The control system is similar in principle to that employed on the Metropolitan trains excepting that on the District Railway the automatic switches—termed contactors—of the individual car controllers are operated electrically instead of pneumatically. At each end of the train there is a master controller in the driver's compartment, by which all the motors are controlled simultaneously.

The signalling system on the District Railway is unique for this country, being operated automatically by the trains themselves through the agency of electro-pneumatic motors attached to each signal post. It is sought by this system—which is owned by the Westinghouse Brake Co.— to eliminate as far as possible the " human element," which has been such a fruitful source of railway accident. It will not be necessary to enter into minute details of the equipment; broadly, the system operates upon the following lines, illustrated diagrammatically in Fig. 53. The track rails play an important part, conducting the current which controls the signal motors. One of the rails is electrically continuous throughout, but the other is cut up into block sections of a length corresponding to the positions of the signals. In each substation is a small motor-generator

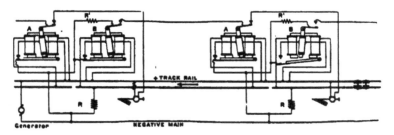

FIG. 53.—DIAGRAM OF WESTINGHOUSE AUTOMATIC TRAIN SIG-
NALLING SYSTEM, DISTRICT RAILWAY.

set which generates direct current at 70 volts pressure, and the + pole of this is connected to the continuous rail while the − pole is connected to an insulated feeder cable, joined up electrically to each section of the other running rail. At the end of each block section are two relays or automatic switches actuated by the motor-generator current in the rails, and by means of a subsidiary circuit these control the valve of the pneumatic motor which moves the signal arm. When there is no train in the block section, current flows from the motor-generator along the continuous rail, through the coils of the relays, back to the sectioned rail, and so to the motor-generator by the insulated feeder. Under these conditions the circuit which controls the valve of the pneumatic signal motor and keeps it open is completed, and the motor holds the arm at "clear" against the pull of gravity. But when a train enters the block section, the motor-generator current flows by preference through the wheels and axles instead of going round the relay coils, which are of comparatively high resistance. The relays now automatically open the circuit controlling the air motor valve, which closes, putting the motor out of action and allowing the signal to move to "danger" by gravity. So long as the train is in this section the signal in its rear will stand at "danger"; but directly the train leaves, the current flows again through the relay coils, with the result that the air motor becomes operative and brings the signal to "clear." The compressed air at 80 lbs. pressure, for actuating the signal arms, is supplied by small motor driven compressors situated in the substations.

The Westinghouse is not the only system of automatic electrical signalling using the running rails, or "track circuit" as it is termed. But it is one of the very few which may fairly claim to be free from the usual objection to such systems, that is, interference with the operation of the signals due to stray currents finding their way to the rails. With any method of signalling involving the use of the track as an electrical circuit it is of course out of the question to utilise the rails as a return for the motor current as in tramway practice. For this reason a second conductor rail is employed specially for the return circuit, as will be seen in the following chapter.

An interesting point about the Westinghouse system of automatic signalling is the Train Stop, which works in conjunction with every signal. It is situated in the road bed adjacent to the signal, and consists of an arm which is raised or lowered pneumatically according to whether the signal is at "danger" or "clear." Its function is automatically to apply the brakes to a train which passes the signal when at danger by striking the handle of a cock on the train as it passes. It is thus a safeguard against carelessness on the part of the motor man, and considerably decreases the risk of collision by making it impossible to run past a signal placed at danger without the driver being made aware that he is doing so by a drastic application of his brakes. In regard to this system generally, it might reasonably be anticipated that variations of the electrical resistance between the rails such as would be caused by heavy rain or a flood would materially affect its proper working. This, however, has been found not to be the case, and at the time of writing several months' practical experience under all working conditions has been obtained. It should be mentioned that before its installation throughout the District line it was thoroughly tested in everyday working on a comparatively unimportant piece of line at Harrow.

CHAPTER XXI.

ELECTRIC RAILWAYS.

Substation Equipment—Feeders and Methods of Laying—Metro
politan Railway Substations—Transformers—Dissipation of Heat
Oil Insulation—Efficiency of Transformers—Rotary Converters
Methods of Starting—'Bus Bars—Oil Switches—Switchboard—
Central London Railway Substations.

THE substation equipment of all direct current tramway
and railway systems working upon the principle of high
tension three-phase transmission with transformation to
500 volts direct current, follows certain general lines. These
are well exemplified by the substations of the Metropolitan
Railway. Like the District substations they were fitted
out by the British Westinghouse Electric and Manufacturing
Co., and are good examples of transforming and converting
stations of their class. The principal operations carried
on are two in number; the transformation of the 11,000
volt three-phase current to low pressure three-phase at
approximately 370 volts by stationary transformers, and
the conversion of this to direct current at 500—600 volts
by rotary converters. Before dealing with details it will
be of interest to take a general view of the substation
system as installed on the Metropolitan Railway. In the
previous chapter it was stated that the line possesses its
own generating station situated at Neasden, roughly, six
miles from Baker Street. Three-phase power is generated

at 11,000 volts as at Chelsea, and is distributed to a total of nine substations placed at various points on the line. The feeder cables are of the three core, paper insulated, lead covered, and armoured type; laid for part of their length in wooden troughs filled in solid with bitumen

FIG. 54.—FINCHLEY ROAD SUBSTATION, METROPOLITAN RAILWAY.

compound. For the remaining part where they run through the tunnels of the St. John's Wood line and the Inner Circle, the cables are attached to the walls by brackets. The total length of feeder cable laid in this manner is about seventy-five miles, and the sectional area of each of the cores varies from 0·1 square inch in the smallest to 0·25 square inch in the heaviest cables. The situations of the substations have been chosen with a view to obtaining economical distribution combined with efficient working of the line, at the lowest cost for plant, mains, and buildings. For instance, where the traffic is light they are placed from three to five miles apart as at Neasden, Harrow, and Ruislip. But at the London end, where the traffic is more dense, as at Baker Street, Praed Street, and Euston Road, the substations are comparatively close together, being within a mile or mile and a half of each other. The size of their equipment varies with the part of the line they serve, but with this exception it is similar throughout all the substations.

The 11,000-volt cables enter the building on the ground level, proceeding along ducts to their metering and controlling gear, and thence to the transformers, which as before stated convert the pressure to approximately 370 volts. Referring to the illustration of Finchley Road substation, Fig. 54, the transformers—varying in number—are placed each in a separate brick chamber situated under the switch gallery, and large holes fitted with iron chequer plates are provided for their removal if at any time necessary. This is further facilitated by a hand operated travelling crane, seen at the far end of the room. Each cell is thoroughly ventilated by a large brick shaft common to all, such an arrangement being necessary to dissipate the heat given off by the transformers, which would otherwise become dangerously hot. The transformers are of 300 k.w. and 435 k.w. capacity each according to the size of the substation, and their efficient insulation is provided for by enclosing them in a ribbed iron case filled with a heavy oil. The purpose of the ribs is to assist the dissipation of heat by the provision of a maximum area of contact with the air. With the same view of keeping the transformers as cool as possible in every part, special ducts are provided between the coils and in the iron core, through which the oil circulates, resulting in the absence of excessive heat in any part, however central. The oil is a good conductor of heat and circulates through the passages of its own accord, conveying the heat to the

FIG. 55.—ARMATURE AND STARTING MOTOR ROTOR OF 400 K.W. ROTARY CONVERTER.

outer radiating surface. The efficiency of this class of transformer is high, standing at 98 per cent. at full load and falling off only to 96·1 per cent. at quarter load.

The next step in the series of transformation brings us to the Rotary Converter, a somewhat complex machine which converts the three-phase 370-volt alternating current into 500—600-volt direct current. Three of these sets are shown in Fig. 54, and it will be seen that outwardly they resemble an ordinary direct current multipolar motor or generator.

There is the usual multipolar field magnet, shunt or compound wound, usually the latter, and an armature, commutator, and brushes of ordinary design. The chief outward difference lies in the three slip rings placed on the shaft at the other side of the armature, and in the small polyphase induction motor fixed on an extension of the shaft, Fig. 55. The slip rings—insulated from one another and from the shaft—are connected to properly chosen points in the armature winding, and by this means the 370-volt three - phase current derived from the secondaries of the stationary transformers is led into the armature, causing it to rotate. The action of the machine is such as to convert the alternating power into direct current at approximately 500 volts, which is collected in the ordinary way by the brushes bearing on the commutator at the other side of the armature. No piece of electrical machinery affords a more interesting study than the rotary converter, but its theory is complicated and beyond the scope of anything but a highly technical treatise. For instance, the fact that there is only one winding upon the armature, although it is dealing at one and the same time with two totally different classes of current, is difficult to understand, and practically impossible to explain without entering deeply into the theory of the machine. So far as we are concerned here this and certain other points must be taken for granted. If a complete understanding of the machine be desired it will be necessary to refer to an advanced text book such as Dr. Sylvanus P. Thompson's "Dynamo Electric Machinery," which deals fully with the subject.

The purpose of the small induction motor is to start the

machine and run it up to the requisite speed before connecting the armature to the alternating current supply from the
transformers. The motoring action of a rotary converter is
essentially synchronous, and before it can be switched in on
the alternating current side it must be running at exactly
the same periodicity. In this it is similar to an alternating
current generator, which, as before explained, has to be
synchronised before being switched in parallel with other
machines. The induction motor and its switchgear are
designed specially for this work. It brings the armature of
the rotary converter gradually up to speed, and the final
delicate adjustments necessary for exact synchronous
running are obtained by varying a resistance inserted in
the motor circuit of the motor. For this purpose a supply
of current is taken from the secondaries of the main
transformers. Rotary converters may also be started
and synchronised from the direct current side if a supply
of direct current is always available. In this case they
are run up as shunt wound motors, synchronism being
attained by varying the resistance in the field circuit.
The second method of starting is efficient in every way,
but for complete safety it calls for a stand by supply of
direct ‚current‚ to meet the possibility of a general shut
down of the whole system. Where this method of starting
is employed it is usual to instal a small battery of
accumulators, or a motor-generator taking current on the
motor side from the alternating circuit and generating
direct current.

It would of course be possible to utilise motor-generators
in traction substations instead of rotary converters, and this
has frequently been done. It is not, however, considered

the best practice for reasons of cost and all round efficiency. The rotary converter is economical at all loads, its transformation efficiency at full load being particularly high. In the early days it was a troublesome machine to operate because it was not properly designed and its action was not fully understood. Now-a-days practically all the initial difficulties have been overcome, and there are thousands of horse-power of rotary converters running on traction systems all over the world with scarcely any trouble worth speaking of.

Reverting to the Finchley Road substation : behind the switchboard are the high tension 'bus bars, arranged in brick cells as at Chelsea. Beneath these on the ground level are the oil switches, built by the British Westinghouse Company, and of a different type to those previously described. When these switches are used in connection with a remote control system as at the Neasden generating station they are operated by a solenoid arrangement instead of by the small motor used by the British Thomson-Houston Company. In the Metropolitan Railway substations, however, the system of remote control is not employed, the oil switches being actuated through the medium of mechanical gearing, of which the levers are seen on the edge of the switch gallery. This gear is used for completing circuit only ; the switches are opened by an electrical tripping arrangement in connection with which a push is installed both on the switchboard and on the levers. The switchboard shown in the illustration Fig. 54 contains apparatus for measuring and controlling the whole intake and output of the substation. The rotary converters are started and synchronised by apparatus contained on the first five panels. The starting switches for the induction motors are placed at the bottom of the board, and above these are the three single-pole knife switches controlling the alternating current side, and the necessary ammeters and voltmeters. The direct-current side of the rotaries is controlled by the apparatus on the next four panels. The field regulator is at the bottom, and the main circuit contains the knife switches just above them in one pole, and the circuit breakers at the top in the other. The end panels are for the 11,000-volt feeders and the cables running to the track. There remains little more to be said about the substations, which are essentially simple in their design; we may therefore pass in the next chapter to the track and rolling stock. Before doing so, however, it should be mentioned that the average large tramway system substation is smaller than those of the Metropolitan Railway, although containing similar plant and arrangements. On the other hand, many of the district substations are equipped with machinery of greater capacity. In the case of the Central London Railway the transforming and converting plant is placed at the bottom of the lift shafts, and each substation contains two 900 k.w. British Thomson-Houston rotary converters. Transmission from the power station at Shepherd's Bush is effected by a three-phase current at a pressure of 5,000 volts. This is transformed down and converted to direct current at 500 volts, at which pressure the third rail is supplied.

CHAPTER XXII.

ELECTRIC RAILWAYS.

Metropolitan Railway Track—Live Rail Protection—Central London Railway Track—Rolling Stock—Corridor *versus* Compartment Carriages—Metropolitan Railway Rolling Stock—Motor Equipment—Collecting Shoes—Cable Work—"All Steel" Cars—Metropolitan Railway Control Gear—Metropolitan Railway Locomotives—Central London Railway Locomotives—Steam Train Conversion—Petrol-Electric Autocars on the North-Eastern Railway.

A SECTION of the Metropolitan Railway track showing the running and conductor rails is given in Fig. 56. The former rails weigh 86 lbs. to the yard, and are of the bull head type adopted by the company, spaced to the usual track gauge of 4 feet 8½ inches. Their fixing to the sleepers follows general practice, iron chairs being employed in conjunction with hardwood wedges. Interest centres chiefly in the conductor rails, of which there are two to each track. That on the outside to the left is the "live" rail, and is connected to the + pole of the track feeders: the rail in the centre is the return conductor, being connected to the − pole. Both rails are of the same sectional area and shape, and their weight is 100 lbs. per yard. They are made from a special high conductivity steel in order to keep the loss of power as low as possible, and their electrical resistance is only six and a half times that of pure copper. The "live" rail is insulated from earth by vitrified porcelain insulators fixed to the sleepers by aid of tripod feet. Two clips bearing on the upper side of the flange and connected together by a bolt secure the rail to the body of the insulator, and this in turn is fixed to an extension of the tripod base by a suitable grouting in the usual way. In renewing a rail it is only necessary to cast the clips loose, when the rail may be readily lifted out. The danger to life of a long length of bare conductor charged to a potential of 500 volts is minimised as much as possible by the two planks fixed one on each side for its whole length. Ever since the third rail system has been

FIG. 56.—SECTION OF METROPOLITAN RAILWAY TRACK

in use a great amount of ingenuity has been brought to bear upon the design of protective methods which would not interfere with the proper performance of its function. Systems have been devised which provide also for a cover to the top of the rail, making it, for this reason, much safer than one which only protects the sides. But in practice it has been found that all such arrangements seriously militate against convenient working, and although the system of two side boards cannot be deemed thoroughly satisfactory from the point of view of safety, it is the one which has been proved to meet all-round requirements the best. Even with this arrangement some little trouble was experienced when the line was first opened with the collecting shoes fouling the boards and breaking them away; but this has been entirely got over.   The return rail in the centre of the track is also insulated from earth by a very similar type of insulator.   It is not protected in any way as it is practically at earth potential, and therefore may be safely handled.   Both rails are bonded at the joints— occurring generally every 45 feet—by specially designed copper bonds fixed to the underneath side of the flanges. To afford the area of contact with the rail necessary to carry the heavy current three bonds are employed at each joint, being attached in the usual manner by hydraulically driven drift pins.   The bonds are built up of several pieces of flat copper ribbon, the ends of which are fused to heavy copper terminals which make contact with the rails.

The conductor rail construction of the District Railway follows the same lines as the Metropolitan.   The Central London Railway track, however, is different, as here only one conductor rail is employed, being placed in the centre of the track.   The return circuit is formed by the running rails, which—as in tramway practice—are bonded together. They are of channel section, laid on a longitudinal wooden track instead of cross-sleepers, and the bonding is made on the upper side of the bottom flanges.   The conductor rail, of an inverted **U** shape, is of high conductivity mild steel, weighing 85 lbs. per yard.   It is supported on a simple form of stoneware insulator every 7½ feet and is bonded on each side with similar bonds to those used on the track rails.   A very interesting point about the Central London Railway is the manner in which the line is graded so as to place every station on the top of a small hill.   A great economy is effected by this means, as when the train approaches a station it is met with an up grade of 1·66 per cent., which materially reduces its speed and saves wear of the brakes. In the converse manner, when leaving a station the train runs down an incline of 3·3 per cent., rendering the starting easy with consequent high acceleration, and reducing the demand for current by 50 per cent.

Much of the success of an electric railway depends upon the design and construction of the rolling stock and its equipment. The conditions of high acceleration and speed, unknown to steam lines, result in much greater strains on every part of the trucks and bodies, and unless the soundest construction is used throughout, the cost of upkeep is liable to become excessive. There is no gainsaying the fact that the public expect much more in the way of comfort and decoration from a new electric line than that which they have become accustomed to with steam traction; and it is therefore necessary, in the vital interests of popularity, to equip an electric railway with rolling stock of a far superior class. The American type of corridor carriage, which for the time appears to have become the standard in this country, was adopted for various reasons. It is more roomy than the British type of compartment carriage, gives free passage the entire length of the train, adapts itself perhaps more readily to decorative and lighting effect, and was found to be popular and successful in America. On the other hand, experience has brought to light some qualities which are by no means appreciated in this country, and it is no exaggeration to say that, generally, it is not so well liked as the older and purely British type. The corridor car is apt to be draughty and cold in winter;

for British tastes it affords too much facility for overcrowding; it does not possess the same seating capacity for an equal length of train; and it is, of course, entirely devoid of that privacy, or comparative privacy, so dear to the insular heart. These are objections which might be overcome, respectively, by a modified design and by the gradual educational process arising out of continued use; but there are other and more vital failings about the corridor car which in the minds of some experts have assigned to it a short life on British lines. A compartment train of the greatest length requires only two guards, the doors being attended to either by the passengers themselves or by station porters. A seven-coach corridor train calls for no fewer than six gatemen, and in the case of every train of this size the wages therefore go up 300 per cent.: a very serious increase and one without adequate compensation in convenience or any other direction. The District Railway has already felt the pinch of increased wages, and there are not wanting many who stigmatise the introduction of corridor cars on this line as little short of a blunder. There is, however, a still further objection of considerable weight where frequent schedule City traffic is concerned, and that is the time taken to load and unload the trains. In this direction the corridor car, with its two combined exits and entrances at the ends, is not comparable to the compartment carriage, as all who have witnessed a City rush at the busy hours must agree. The District Railway tried to get over the difficulty by the introduction of a centre door, as used in America; but so far the British public has declined to be coerced into using one for an exit and the other for an entrance, with the natural result that the centre door as a means for facilitating combined entrance and exit has failed lamentably. The end of the whole question is as yet a matter of doubt; and, many companies being so heavily committed to the corridor type of carriage, it is extremely unlikely that changes will be

FIG. 57.—METROPOLITAN RAILWAY TRUCK WITH MOTORS IN POSITION.

made for some years. It is a significant fact, however, that one of the most up-to-date American lines, the Illinois Central, has in part reverted to the compartment carriage, with a reduced wages bill and increased convenience; and when with this is coupled the opinion of numerous traction experts, to the effect that the corridor car is unsuitable for City service, it would appear that there is at least a reasonable chance of the old compartment carriage, so

R 2

FIG. 58. CURRENT COLLECTING SHOES, METROPOLITAN RAILWAY.

ruthlessly thrown aside, resuming its place on British lines, although doubtless in an improved form.

The corridor rolling stock of the Metropolitan Railway is an excellent example of first class work. It was made to British designs by the Metropolitan Amalgamated Railway Carriage and Wagon Co., Ltd., of Birmingham and Manchester, and its construction is marked by that sturdiness characteristic of British work. Each car is mounted upon two four-wheel bogie trucks, which in the case of the motor cars carry two 150 h.-p. railway motors of British Westinghouse manufacture on each. There are only two motor cars to each train instead of three, as on the District, and they are situated one at each end. Thus, a train of ordinary length carries 1,200 h.-p. of motors; that is, 300 h.-p. on each of the four trucks. One of these, with the motors in position, is

FIG. 59.—METHOD OF CARRYING OUT CABLE WORK ON METROPOLITAN
TRAINS.

shown in Fig. 57. The equipment as described above is
capable of propelling a train of 140 tons weight at

a speed of forty miles per hour on a straight and level track ; and from the point of view of rapid acceleration its performance is very satisfactory and considerably superior to that of the old steam locomotives. The current-collecting gear is shown in Fig. 58. It is very simple in design, consisting only of a shoe which makes rubbing contact with the conductor rail, pressed down by a helical spring. It will be noticed that the shoe is suspended by slotted links at each end to give it free vertical movement, and that two flexible copper strips are provided to conduct the current away from the rubbing surface. The equipment, as shown, is carried on a beam, and consists of three shoes. The two outer ones are for making contact with the " live " rail whichever side of the train it happens to be on, and the inner one, of which the top only is visible, is for returning the current to the — or centre rail. One of these sets is fixed at each end of the motor cars, and in addition there is a further return shoe which is placed by itself in the centre of the coach.

Fig. 59 illustrates the manner in which the cable work on the car is carried out. To minimise the risk of fire the wires are laid in asbestos grooves filled in with a composition of a fireproof nature and covered with sheet asbestos. The floors are of sheet steel surfaced with a hard fireproof material five-eighths of an inch thick. In every detail of electrical rolling stock the most thorough precautions against fire are necessary. In the case of tube lines particularly the tunnel acts as a powerful draught-producing flue directly fire is started, and to minimise the chance of the fire obtaining a hold, all wood-work is treated with a fireproofing composition, and many of the cables are

insulated by asbestos. What is termed the " all steel " car
has of late come to the front for the double reason of comparative freedom from fire risk and reduced weight for an equal

size of coach. The saving in dead weight is approximately
five tons, and it is estimated that this results in a monetary
saving per annum of about £175 per car, owing to the smaller
amount of power required.

The control gear of the Metropolitan trains is extremely
interesting, but complicated and somewhat difficult to follow
if its full details be gone into. For the purpose of obtaining
a grasp of its general principles it will only be necessary to
consider discursively the two chief parts, the turret controller and the master controller. The former takes its
name from its shape, and is shown in position close to the
truck in Fig. 60. It consists of fifteen pneumatically
operated switches grouped radially round a central air
chamber and magnetic blow-out coil, the former on top of
the latter. The pneumatic motors are simply small air
cylinders containing a piston, which, working against a
powerful spring, closes the switch to which it is attached.
Opening of the switch is performed by the spring directly
the air pressure is removed. The pneumatic cylinders take
their supply of compressed air from the central chamber—
fed by a compressor—and their operation is controlled by
an electrically actuated pin valve. When current is passed
through the coil of the valve magnet it opens the port,
allowing air to pass to the cylinder from the central chamber
and thus closing the switch. Directly current is cut off
the valve magnet, the air pressure is removed and the
switch resumes its natural open position. The switches
are placed at the base of the turret, each in a rectangular
fireproof compartment of vulcabeston by itself. Between
each of these compartments a U-shaped piece of iron will
be noticed, pointing alternately up and down. These

FIG. 61.—MASTER CONTROLLER AND AIR-BRAKE GEAR IN DRIVER'S COMPARTMENT, METROPOLITAN RAILWAY.

are the pole-pieces of the central magnetic blow-out coil, and their arrangement in this manner provides a powerful field which passes through the jaws of the switch, deflecting the arc formed at opening or, in other words, blowing it out. The closing and opening of the switches in their right order produces substantially the same effect as the movement of a tramway controller handle. The motors are first placed in series together with resistance, which is gradually cut out; they are then connected in parallel, also with resistance in circuit, and finally they are switched in full parallel each across the 500 volts when the train is running at full speed. The resistances are shown to the left of the turret controller in Fig. 60. They are built up of cast grids bolted together in an iron frame. The master controller, with the handle removed, is shown in Fig. 61, fixed in position in the driver's compartment to the right of the air brake gear. It is remarkably small and compact, as it is only called upon to operate with a 14-volt current obtained from a small storage battery. It is this current which actuates the pin valves of the turret controller pneumatic pistons, together with all other auxiliary controlling gear. There are ten positions, or notches, in the master controller: five being for forward and five for backward running. At the first, the emergency brake is magnetically made ready for action. At the second, what is termed the reverser —a switch that governs the direction of running—is closed, in conjunction with the main supply circuit. Movement to the third notch puts the motors in series together with all the resistance. The fourth notch brings an automatic arrangement into play which cuts out the resistance step by step. At the fifth position all the switches previously closed are opened, and simultaneously another set are closed which place the motors in parallel with resistance in circuit. This is then automatically cut out as before until the motors are in full parallel. To prevent undue rushes of current consequent upon careless use of the controller an ingenious device termed a limit switch is installed. The function of this switch is to preserve an uniform accelerating current, which it does automatically by not permitting the controller switches to close and cut out resistance unless the current passing through the motors is within the proper limits. Owing to the presence of this switch it would do no harm to throw the controller handle suddenly over to full parallel. In such a case the turret controller would go through all its various positions automatically and under the control of the limit switch. Automatic acceleration of this kind has been found to result in a saving in current of from 10 to 15 per cent. over hand control, the reason being that the gear is set to operate in the most economical manner, from which it cannot depart, whereas, on the other hand, the control is left to the judgment of the driver. The handle of the master controller works against a spring, and should the motor man leave go, it would fly back to the neutral point between forward and backward running, and at the same time automatically apply the brakes through the agency of a small subsidiary circuit controlling them magnetically. This is a very necessary safeguard where everything depends upon one man, as is the case in most electrical railways.

In addition to their motor car rolling stock the Metropolitan Railway possess a number of electrical locomotives used for main line traffic between Baker Street and Harrow, at which latter point steam locomotives pick up the train and continue the journey to Aylesbury, etc. They

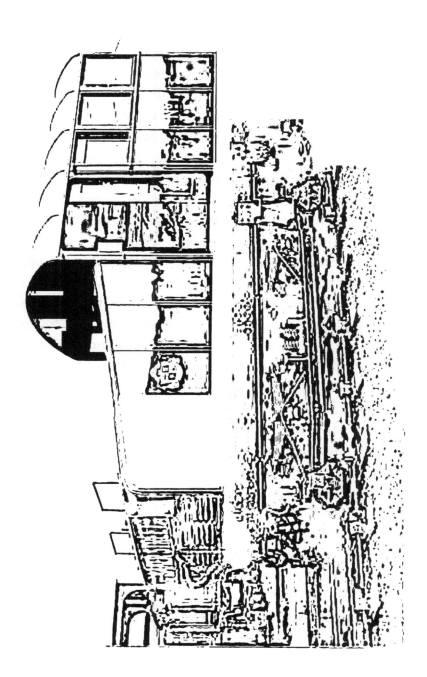

are also employed on the inner circle for hauling the Great Western trains from Bishop's Road, and for goods traffic. One of these locomotives is illustrated in Fig. 62. Its weight is 50 tons, and its four-motor 1,200 h.-p. equipment will propel a 120-ton train at a speed of 36 miles per hour on the level. An interesting point about these locomotives is that their motors are artificially cooled by air jets. Owing to restrictions of space at the termini it was necessary that the locomotives should be as short as possible, and it was found impracticable to employ motors of the correct rated capacity for the reason that they took up too much room. The alternative was to use motors of lower horse-power, overloading them, and keeping them cool by forced ventilation. This arrangement has proved to be most satisfactory, no trouble having been experienced.

The locomotives originally employed on the Central London Railway, and now superseded by the multiple unit system were of much smaller capacity than the above. The equipment consisted of four 117-h.-p. motors of the British Thomson-Houston Co's. make, and the weight of the locomotives was 43 tons. So far as haulage was concerned they were found quite satisfactory, and their replacement by the multiple unit system was due solely to vibration troubles occasioned by their weight. The splitting up of the driving power by the substitution of a motor car at each end of the train has resulted in a material diminution of the hammering which originally took place at the rail joints.

A recent interesting development on the Metropolitan Railway is the conversion of a couple of trains designed for steam haulage to electric traction. One of the most serious financial difficulties with which a steam line is faced when changing to electrical propulsion is the scrapping of existing rolling stock, and, therefore, the means taken by the Metropolitan Company to avoid this where possible are of interest. The company was in possession of several comparatively new trains designed for steam haulage, and two of these were taken and adapted in the following manner. Little alteration has been made in the trailing cars beyond the substitution of air for vacuum brakes, but the coaches which formerly contained the luggage and guard's compartment have been fitted with new bogie trucks of a stronger construction, and suitable for accommodating the motors. Of these there are four, of 200 h.-p., to each motor coach, and the total of 1,600 h.-p. per train is found sufficient to attain a speed of 45 miles per hour on the level. The motors were supplied by the British Thomson-Houston Co., who carried out the whole of the conversion. The controlling gear is placed in the guard's compartment, now appropriated by the motor man. It provides for the usual series-parallel control of the motors, the method being similar to that employed on the District Railway, and known as the Sprague Thomson-Houston multiple unit system. These trains have now worked for several months without a hitch, and the experiment is regarded as an unqualified success.

The petrol-electric autocars recently put into service on the North Eastern Railway constitute a most interesting departure from usual practice. They are designed to secure the advantages of electric traction on a single-car suburban service too small to warrant the installation of the usual generating station and conductor rails; and several months' working under practical conditions have shown very satisfactory results. Each car carries its own generating plant in the form of an 80-h.-p. four-cylinder horizontal petrol engine running at a normal speed of 420 revolutions per minute, coupled direct to a separately excited compound wound direct current dynamo of 55 k.w. capacity. This generator is designed to work at any pressure between 300 and 550 volts by variation of the current in its shunt field circuit, which is excited by a 3·75-k.w. 72-volt, shunt wound, direct current machine, driven by belt from the main shaft. The generating equipment also includes a 35-cell storage battery. The car is fitted with two 55-h.-p. series wound railway motors, which are controlled on the series-parallel system in the ordinary manner. The working of the equipment is briefly as follows. At the commencement of a run the generator is operated as a motor from the battery, and by this means the petrol engine is started up. When it is running, the generator is disconnected from the battery, and upon a pressure of approximately 400 volts being reached, current is switched on to the motors through the controller. The pressure is then gradually increased to 500 volts by cutting out resistance in the shunt circuit, and the car accelerates up to a speed of 36 miles per hour. In stopping, current is switched off the motors and a magnetic track brake brought into play by moving the controller over to the braking notches. The engine is left running, and when the voltage of the generator has been reduced back to the original pressure of 400 volts, everything is in readiness for starting again. The weight of the car and its equipment is 35 tons, and it possesses seating capacity for fifty-two passengers. In addition to its ordinary function the exciter is used for lighting the car, and, when this is not required, for charging the battery.

Railway motor cars of various kinds are becoming very popular for light suburban service, and large developments in this direction may be anticipated, as their operation when properly designed is cheap and satisfactory.

CHAPTER XXIII.

ELECTRIC RAILWAYS.

FOR several years alternating current electric traction received attention in Europe alone, the great electrical firms of Germany and Switzerland being well to the front in experimental work made with a view to adapting the three-phase motor to the requirements of railway traction. America lagged behind, and has only recently come forward with the single-phase alternating current motor, which to all appearance has a great future before it.

The limitations of direct current as regards pressure and efficiency of transformation were seen at an early date to preclude it from ever becoming the source of power for main line or any but suburban railways. If it were possible to generate direct current at several thousand volts, and use it at that pressure in the train motors in conjunction with

an overhead wire, less would be heard of alternating
current traction, particularly of the system employing
three-phase motors.   But, owing to commutation difficulties,
it is not practically possible, at any rate at the present
date, and the only means available are those employed on
the District Railway, for instance, the capital and running
cost of which would be prohibitive for lines of a much lower
traffic density.   One essential of main line operation is that
if pressure transformation be necessary it must be carried
out by the simplest and most efficient means and without
the aid of costly rotary machinery requiring attention.
Further, it is desirable that the number of sub-stations
should be small, which directly points to the use of high
pressure current on the train in conjunction with overhead
wires.   Obviously these conditions can only be fulfilled by
an alternating current system, in which there is no
difficulty in generation and transmission at high pressure,
and a maximum of efficiency in transformation either at
sub-stations or on the train.   At the time the subject first
received attention there was no question of employing
single-phase motors, which were then quite impracticable
for the purpose.   The three-phase motor, on the other hand,
even at that date possessed characteristics which made it
worthy of consideration for traction work, and a few
Continental firms, notably Messrs. Ganz, of Buda-Pesth,
undertook a series of experiments with encouraging results.
In addition to the inherent economic qualities, such as
transformation, possessed by all alternating current systems,
the three-phase motor itself is in some respects a good
power agent for traction purposes.   It is mechanical, and
calculated to withstand rough usuage, and it will return

power to the line when running downhill over a certain speed and exercise a very powerful braking effect. Against these good points must be set down several disadvantages, some inherent to alternating current traction generally, and others peculiar to the three-phase system. The most pronounced defects of the former lie in the somewhat larger feeders required as compared with direct current, owing to

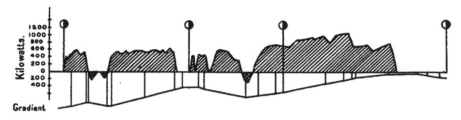

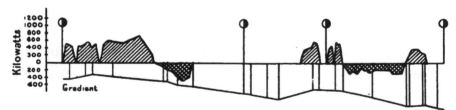

FIG. 63.—VALTELINA THREE-PHASE RAILWAY. DIAGRAM SHOWING POWER ABSORBED AND REGENERATED.

certain causes which need not be entered into ; the greater pressure drop in the rails when they form part of the circuit, due to the fact that being of steel they produce a highly inductive effect ; and insulation difficulties with the overhead wire when it is supplied at high pressure. The three-phase motor suffers from the following disadvantages · It is comparatively uneconomical in speed control, and does not accelerate well; for efficiency there must be a very small clearance between the rotor and the stator, which is liable to give rise to trouble; and it requires at least **two** overhead wires, which introduce considerable difficulties **at** crowded intersections. Taking these good and bad **points** in further detail: The manner in which the polyphase motor will return power to the line is very striking and to a large extent valuable. It reduces the wear on brakes, and materially affects in a favourable manner the power consumption of the whole system. This is forcibly illustrated in Fig. 63, which refers to the Valtelina line, and is adapted from a diagram published in *Le Génie Civil.* The upper half of the figure deals with the outward journey and the lower with the return, the distance being 6·8 miles over a hilly track. The hatched curves above the line represent ·the power taken in k.w.; and the double hatched portions below the line show the amount regenerated. The curve underneath in each case represents the contour of the track, giving an approximate idea of the various gradients. The black and white circles are stopping places. It will be noticed that the train starts on an up grade, and takes power to the average extent of about 450 k.w. A little way out, the line descends, and the motors return about 200 k.w. for a short period. A long up grade is then encountered, and at its summit is a station, into which the train runs by its own momentum, power being shut off just previous to reaching the top of the ascent. The next portion of the curve shows the current taken at starting, and although there is now a down grade, no power is returned to the line for the greater part of it because in this particular test the train was here run up to full speed for the first time, and the conditions under which the motors were operating were somewhat different to those which obtained during the early part of the run. However, a little is returned just at the foot of the descent when the highest speed has been reached ; and now, as the line is uphill most of the way, power is absorbed for the rest of the run, with the exception of the last mile which is half level and half down grade and consequently permits of current being cut off the motors some time prior to stopping. Compared with the outward journey, the return is practically all downhill. The start on a slight up grade absorbs power, as does the acceleration up to full speed on the first down grade. But, as this gets more pronounced, regeneration takes place to the extent of about 400 k.w. for part of the way, the train, after this has ceased, running into the first station by its own momentum. The start downhill absorbs no power, and the greater part of the succeeding up grade is negotiated by the momentum. After starting again, a long period of regeneration takes place, power being applied only for about a quarter of a mile before the first stop. It will be obvious that the amount of current thus returned to the line is valuable, as it assists other trains requiring power at the time regeneration takes place, and correspondingly reduces the load on the generating plant. The one disadvantage to this process lies in the fact that the motors must be somewhat heavier and more costly than where regeneration is not employed, since instead of having their periods of rest when running downhill, they are at work the greater part of the time, either motoring or generating. In this particular test the train weighed 286 tons on the outward, and 280 tons on the return journey. The two running speeds were respectively 19 and 38 miles per hour.

To turn now to the disadvantages of the three-phase motor. It is inherently a constant speed machine, and therefore not at all well adapted to traction work, of which continual speed variation is often a characteristic. As explained in Chapter VI., it is always possible to alter the speed of a polyphase motor through a wide range by varying the value of a resistance placed in the motor circuit, but this is inefficient in an increasing degree as the speed is lowered; for instance, if it be reduced to 20 per cent. of the maximum and only economical value, the efficiency of the motor will drop to 20 per cent. There is a way of mitigating this difficulty, and it is employed on all polyphase lines. The equipment, which ordinarily would consist of two motors, is increased to four, and to obtain an economical low speed, the rotor circuits of two of the motors are completed through the stator windings of the other two. The arrangement is termed " concatenation " or " cascade," and without entering into its theory it will be sufficient to state that it makes running at half speed economical. It suffers, however, from the initial disadvantage that the cost and weight of the equipment are unduly high, and that the second pair of motors are quite useless at above half speed. This tells particularly on long high speed runs, where the power absorbed in carrying the extra weight might possibly exceed the amount saved by the use of the concatenation principle at low speeds. The other important point that militates against the all-round effectiveness of polyphase motors is the overhead wire trouble. If the rails are used as one leg of the three-phase circuit, as in general practice, two trolley wires must be provided for the other legs; and if the rails are not used, three overhead wires must be employed. It is impossible to do with less; and the duplication of the trolley system adds at least 100 per cent. to the difficulty of maintaining efficient insulation, and to the complication at intersections. With a single overhead wire it is only necessary to provide for efficient insulation between it and earth; with two wires this must be done in each case, and, in addition, they must be thoroughly insulated from one another. In view of all the foregoing points it will be seen that the position of the polyphase system is much as follows. It is efficient where the traffic is conducted at practically uniform speed on a simple line with few stops; but it is not suitable where constantly varying speed is necessary, and undesirable where the line equipment calls for complicated overhead work.

It will be of value before leaving the subject to give some data of the principal polyphase railways that have been operated under practical conditions. The first was a light tramway line at Lugano, erected by the Swiss firm of Brown, Boveri & Co., of Baden, in 1895. 5000-volt three-phase current is generated by water power about $7\frac{1}{2}$ miles away, and transformed down to 400 volts, at which pressure the motors are designed to work. Each car—weighing $6\frac{1}{2}$ tons —carries a 20 h.-p. three-phase motor, and runs at a speed of 9 miles per hour. The Valtelina line, 67 miles long, employs a transmission pressure of 20,000 volts and a trolley pressure of 3,000 volts, attaining a speed of 42 miles an hour with a 600 h.-p. locomotive equipment. The highest trolley wire pressure is the 10,000 volts employed on the experimental Berlin Zossen 14 mile line. A motor car equipped with 3,000 h.-p. is used, and the speed it reaches is 102 miles per hour. In this case the pressure is reduced on the train by aid of transformers. The Burgdorf-Thun Railway, opened in 1899, is 25 miles long, and uses both locomotives and motor cars. They are equipped respectively with 300 h.-p. and 240 h.-p. of motors, and the speed of passenger trains is 22½ miles per hour. Power is generated at 4,000 volts, transformed up to 16,000 volts for transmission, and transformed down again to the trolley pressure of 750 volts. Other small polyphase lines exist in the mountainous districts of Switzerland, of which mention need not be made. It is possible that further expansion of the system may take place in the direction of small and unimportant lines, but every indication points to the single-phase system as the one that will practically monopolise attention in the future.

We are here introduced to an entirely new class of motor, which, although known for many years in an experimental form, has only recently taken practical shape. It is well known that a series wound direct current motor will maintain the same direction of rotation irrespective of the direction of flow of current through its windings. To reverse the direction of running it is necessary to reverse the direction of flow in either the armature or field circuit; if both be reversed at the same time, say by changing the polarity of the supply, there is no effect on the direction of running. It follows, therefore, that an ordinary direct current series wound motor will rotate in the usual way if a single phase alternating current be supplied to its terminals. This fact was known many years ago, and small motors were built, which ran, although in a very inefficient and generally unsatisfactory manner. It remained chiefly for Lamme, of the Westinghouse Electric Co.; Steinmetz, of the General Electric Co.; and Finzi, an Italian engineer, to improve matters in this respect, and at the present time there are many single-phase railway motors in operation, built mostly from the designs of these experts, which are proving reliable and efficient in every way.

Although on general lines there is little difference between

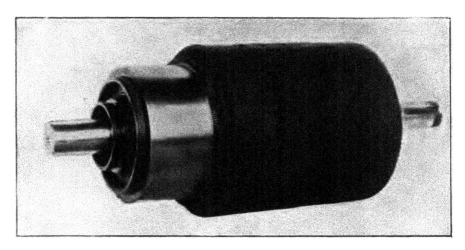

FIG. 64.—ARMATURE OF WESTINGHOUSE SINGLE-PHASE SERIES RAILWAY MOTOR.

the single-phase series motor and its direct current prototype so far as the principles of working go, there are some detailed but highly important actions present in the former which are entirely foreign to the latter. The fact that an alternating and not a direct current flows in the windings makes all the difference in these points. Owing to this, the field magnet core throughout must be built up of laminations to avoid the excessive loss that would otherwise take place by the induction of local currents in the core itself, with consequent heating and inefficiency. In addition, it must be remembered that the presence of an alternating current in the field and armature windings must result in the generation of various E.M.F.'s in the coils which are entirely absent in the direct current machine, and the action of these has to be provided for in the design, and where necessary neutralised or compensated for in order

FIG. 65.—FIELD MAGNETS OF WESTINGHOUSE SINGLE-PHASE SERIES RAILWAY MOTOR.

that efficiency may not be impaired. These are details of a highly technical nature which must be left out of consideration. Though vitally important, they have little influence on the appearance of the motor, and in any case do not enter into the general principle. The armature and field systems of a single-phase railway motor are shown in Figs. 64 and 65. It will be noted that the armature and commutator are indistinguishable from those of a direct current machine ; the field magnets also possess outwardly much the same characteristics as a four pole traction motor. These machines have now been brought to a pitch of perfection at which their running compares well with any direct current traction motor. They possess an excellent starting torque, do not spark at the brushes, and will stand rough usage to any reasonable extent. Their chief value, however, lies in the facts that besides making single-phase electric traction commercially possible, with all the concurrent advantages of efficient transformation and simple overhead work, they will operate equally well with direct current, and when working on the alternating current circuit may be controlled in a more economical manner than the direct current motor. With regard to the first point, it is clear that such a machine presents no difficulty in its operation on a direct current circuit, and this means that in places where a high pressure overhead wire might be objectionable, such as at crowded termini, the ordinary third rail construction may be resorted to with a direct current supply at the usual voltage. Further, it allows rolling stock equipped on these lines to run over other companies' track which perhaps is fed on the ordinary direct current system. Under certain circumstances both these features might assume high importance, as may well be imagined. The question of control is always one of weight, and in this respect the single phase system possesses distinct advantages. It has been customary so far to supply the overhead line at a pressure of several thousand volts, transforming down on the train to some suitable low pressure. This is effected by means of an ordinary alternating current transformer, and by the simple addition of a number of tappings taken off the

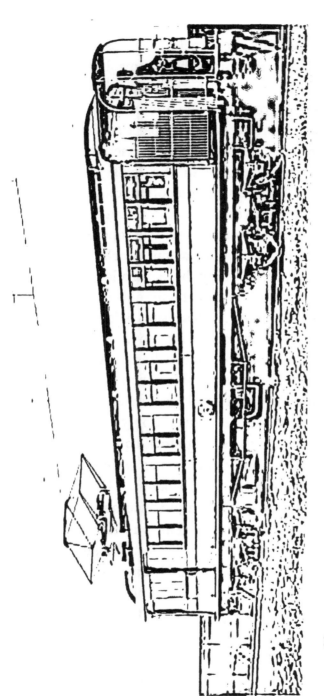

FIG. 66.—COACH FITTED WITH SINGLE-PHASE RAILWAY EQUIPMENT. CATENARY OVERHEAD WIRE SUSPENSION.

secondary winding at different places, it is possible to apply a very wide range of voltage to the terminals of the motors, with the result that their speed is readily controlled without the use of the wasteful resistances found at so many points in the ordinary series-parallel direct current controller. Further, every position, on the alternating controller is a " running " position, for the reason that resistances are entirely absent, and this affords a much wider range of speed and flexibility of working. The series-parallel arrangement can of course be employed, and there are still further methods which have been used ; but the variable ratio transformer, as it is termed, seems to offer the best solution when everything is taken into considera- tion. In the event of rolling stock being destined for use on both alternating and direct current track, it is necessary to instal an alternative series-parallel control, as the transformer is, of course, useless on direct current.

Fig. 66 shows the first car in Great Britain to be equipped with commercial single-phase railway motors. It is now running on a demonstration line at the Trafford Park Works of the British Westinghouse Co., with whose apparatus it is fitted. It will be noticed that the car is lettered " Metropolitan Railway," but this is only because it has been borrowed from that company for the purpose of demonstrating the practicability of the single-phase system with proper rolling stock. The style of collecting gear is that known as the pantograph, and the object of this con- struction is to permit of high speed running and reversal of direction without any corresponding adjustment of the gear. The overhead line has been constructed upon the catenary principle, which in one form or another is the only suitable method to meet the requirements of high speed. It provides an even trolley wire, that is one free from sag, and further, it is a safe and efficient system of suspension for a high pressure conductor. An insulated steel stranded wire is carried from pole to pole and adjusted to a certain degree of sag. From this is suspended the trolley wire by aid of hangers of different lengths, the shortest being used at the centre of the span and the longest near the poles, and the obvious result is a level trolley wire such as is essential for high speed working. This type of overhead construction is applicable to both bracket and span wire suspension, and it is almost invariably used on high pressure overhead conductor railways. The Trafford Park demonstration line is three fifths of a mile long, and the gauge is the usual 4 ft. 8½ inches. The overhead wire is fed with a 25 period alternating current at a pressure of 3,000 volts, and the track rails are bonded to form the return circuit. The car weighs 36 tons empty, and the equipment comprises four 100 h.-p. motors designed for working at the transformed pressure of 250 volts. This practically represents the standard Westinghouse system, but further examples of lines working commercially will be given later.

In the meantime we may turn to an interesting single-phase railway running between Murnau and Ober-Ammergau, a distance of about 14 miles. Its equipment was completed in 1905 by Messrs Siemens-Schuckert, who have carried out some very successful work with a single-phase system of their own. Motor cars are employed with trailers, and each car is fitted with two 80 h.-p. 10 pole series alternating current motors and two variable ratio transformers,

FIG. 6. 1,350 H.-P. SINGLE-PHASE RAILWAY LOCOMOTIVE.

one for either motor. The normal running speed is
25 miles per hour, and the starting, even on a steep

up grade with a heavy load, has been found very satisfactory. The overhead wire is fed direct from the generators at a pressure of 5000 volts and a frequency of 16⅔ cycles per second, and the track is bonded to form the return circuit. The trolley gear on each car consists of two pneumatically operated collectors of bow form, which are held in position by springs and moved when necessary by air motors supplied from the brake compressors. The transformers reduce the line pressure to 260 volts when all the secondary winding is in circuit, and connection with the various tappings is made through a barrel controller fitted with a magnetic blow out to keep down the sparking when passing from stop to stop. Motors of similar construction to those used on this line are built by the British firm of Siemens Bros., Ltd., at their Stafford works.

A very interesting single-phase locomotive built by the Westinghouse Electric Co., of Pittsburgh, U.S.A., is illustrated in Fig. 67. It is stated to be the largest alternating current locomotive in the world, and to be equipped with the largest single-phase series motors yet made. It was built for the special purpose of demonstrating practically the possibilities of heavy single-phase traction, and it has shown some very interesting and favourable results. As will be seen, it is built in halves, each of which is independent of the other, but it is designed to work normally as a single unit. Its weight complete is 135 tons, and it is equipped with six 225 h.-p. motors, one on each axle, two transformers, which reduce the pressure to the 325 volts required by the motors, and two regulators of a different type to those mentioned heretofore. Current of 25 cycles at 6,000 volts is supplied to the overhead wire, whence it is

collected by two pantograph trolleys and passed through oil switch gear to the transformers. These, like the motors, are cooled by compressed air, and the collectors are raised or lowered by the same means, the supply, as usual, being obtained from the brake equipment. By aid of the regulators it is possible to vary the pressure at the motor terminals from 140 volts to the full 325, and any pressure between these two limits is available for speed control without undue loss of efficiency. The maximum speed is about 30 miles per hour, but the locomotive has been designed specially for low speed goods service, and when drawing its full load of approximately 1,200 tons its speed is 10 miles per hour.

One of the most recent single-phase lines is that of the Spokane and Inland Railway Co., Washington, U. S. A. Its equipment was completed towards the end of 1906 by the Westinghouse Electric Co., whose standard single phase system is used. There are in all about 106 miles of line over which both passenger and freight traffic is conducted, motor cars and locomotives being employed. Three-phase 60 cycle power at 4,000 volts is purchased from a local company, and after its conversion to 25 cycle 2,200 volt single-phase in a motor-generator sub-station, it is transformed up to 45,000 volts for transmission to the fifteen sub-stations scattered along the line. Here it is transformed down to the trolley pressures of either 6,600 volts, for use in the open country, or 700 volts in the towns, and in addition, at Spokane there is a 600 volt direct current supply upon which the same rolling stock and motors are used. Each motor car equipment consists of four 100 h.-p. series motors, capable of attaining a speed of 40 miles per hour, and control is arranged for on the series-parallel system in order that to a large extent the same gear may be used on alternating and direct current. The locomotives are equipped with four 150 h.-p. motors, which will maintain a speed of 30 miles per hour on level track when drawing a full load. The control is similar to that of the motor cars, being effected by the aid of a series of pneumatically operated switches governed by a master controller in much the same manner as on the direct current Metropolitan trains. Two trolleys are used, one being of the pantograph type, for employment when running on alternating current, and the other of the ordinary tramway description, for service on the direct current section. The substitution of the one for the other is the only operation necessary when passing from the alternating to the direct current supply, or *vice versâ*; the re-arrangement of the electrical gear is done automatically by means of apparatus specially installed. The full engineering details of this line are very interesting, and they comprise many novel features; their technicality, however, in conjunction with considerations of space, renders their omission necessary. The experience gained in one or two years of actual working will be of considerable service in determining the true value of the system; and this line, owing to its rather unusual equipment necessitated by the three pressures at which it operates, will undoubtedly be watched with interest.

In conclusion, it will not be out of place to mention the latest and perhaps most important development in the application of single-phase current to railway traction, viz., the suburban line now being equipped by the London, Brighton, and South-Coast Railway. The conversion from steam traction to the single phase system was decided upon towards the end of 1905, and the contract was placed with the Allgemeine Elektricitats Gesellschaft, who had carried out the successful Spindlersfeld Railway on the same system. Ultimately, a generating station will be erected at New Cross, and will supply current of 25 cycles at 6,000 volts to the overhead line, which will be erected on the double catenary principle. Sliding bow collectors will be used, and instead of the rails being bonded together to form the return circuit, they will each be connected to the outer conductor of an underground concentric cable, the inner core of which will act as a high pressure distributor. It is interesting to note that the ordinary type of corridor carriage, with doors only at each end, will not be employed the compartment principle with sliding side doors operated either by pneumatic power or by hand, having been decided upon as better meeting the requirements of the line. The question of single phase versus other forms of electric traction was most thoroughly gone into by the company's directors and Mr. Philip Dawson, their engineer, and the result was an unqualified approval of the former. Upon the strength of this the company have committed themselves to a scheme which can by no means be looked upon as an experiment, and although this course is perhaps open to criticism upon the score of its apparently daring nature, there is no doubt in the minds of those duly qualified by experience to know, that the company have made a wise and enterprising decision.

GLOSSARY.

ACCUMULATOR.—A chemical cell employed for storing electrical energy.

ALTERNATING CURRENT.—A current which alternates in direction of flow.

AMPERE.—The unit of electrical current.

ARC.—The electric flash occasioned by breaking a current carrying circuit.

ARMATURE.—The portion of a dynamo in which the current is generated. In an ordinary motor the part which rotates.

ARRESTOR.—An apparatus for protecting an electric circuit against lightning or other high pressure discharge.

ASYNCHRONOUS MOTOR.—An alternating current motor, the speed of which is variable and not necessarily synchronous with that of the generator from which the current driving it is derived.

BALANCER.—A machine used in connection with a three-wire system of distribution. Its function is to compensate for difference in load on the two sides of the system.

BOGIE TRUCK.—A truck used with long cars to facilitate the rounding of curves. It swivels on a central pivot independently of the car body.

BOOSTER.—An auxiliary dynamo employed for the purpose of keeping one special branch circuit at a higher or lower pressure than that of the main circuit.

CATENARY OVERHEAD CONSTRUCTION.—The suspension of the trolley wire, by means of a parallel wire or wires to which it is attached by hangers of different lengths at short intervals.

CIRCUIT BREAKER.—A switch that automatically opens circuit when a predetermined excess current flows.

COMMUTATOR.—The device fixed on the shaft of a direct current dynamo which rectifies the alternating currents generated in its armature, and sends them out to the circuit in an uniform direction.

COMPOUND WINDING.—The magnet winding of a compound wound dynamo; i.e., one in which both a shunt and a series circuit is wound on the field magnet. The same winding is used for motors.

CONCATENATION.—The principle used in a system of polyphase railway motor control.

CONDUCTOR.—Any body which conducts electricity.

CONTINUOUS CURRENT.—A current which flows uniformly from the + to the − pole.

CROSSING.—A trolley wire fitting employed where wires cross one another.

CUTOUT.—See Fuse.

DIELECTRIC.—Any insulating substance.

DIRECT CURRENT.—See continuous current.

DYNAMO.—A machine for converting mechanical into electrical power.

EDDY CURRENTS.—Local currents generated within a mass of metal by magnetic induction.

ELECTROSTATIC.—Appertaining to a stationary electric charge as opposed to current electricity.

E.M.F.—Abbreviation of electro-motive-force.

EXCITATION.—The magnetising of a dynamo or motor field magnet by passing current through coils wound round the limbs.

EXCITER.—A direct current generator which supplies current to excite the field magnets, usually, of an alternator.

FIELD CIRCUIT.—The windings of a dynamo or motor field magnet.

FIELD MAGNET.—The magnet which generates the lines of force cut by the armature conductors in their rotation.

FIELD (MAGNETIC).—The lines of magnetic force emanating from the poles of a magnet.

FINGERS.—The contact pieces of a motor controller.

FISHPLATE.—An iron plate applied to each side of a rail joint by bolts to bind the rails together.

FLEXIBLE SUSPENSION.—A method of suspending a trolley wire which leaves it free to move up and down at the point of suspension.

FROG.—An overhead trolley wire fitting employed at places where one or more tracks branch off.

FUSE.—A wire or strip of metal gauged to melt and break the circuit when an excessive current flows.

FORMER.—The frame upon which a coil is wound.

HANGER.—A clip for suspending an overhead trolley wire.

INDUCTION MOTOR.—An alternating current motor which depends for the rotation of its moving part upon currents induced in its windings by the stator field.

KILOWATT.—1,000 watts. The unit by which the size of electrical machinery is specified.

LAMINATIONS.—Flat discs or strips insulated from one another, and used in electrical machinery in the place of solid iron.

MASTER CONTROLLER.—The principal controller, by means of which the subsidiary controllers of a multiple-unit railway system are operated.

MOTOR-GENERATOR.—An electric motor geared to and driving a dynamo; used for transformation purposes.

MULTIPLE-UNIT CONTROL.—A system of train operation employing motors. mounted on two or more cars in the place of a locomotive.

NEGATIVE BOOSTER.—A booster so connected as to reduce the difference of potential between the two ends of a return circuit.

NO-LOAD RELEASE.—An attachment to a motor starting switch that allows it to open circuit when current is cut off the shunt winding of a motor.

OHM.—The unit of resistance to the flow of electricity.

OHM's LAW.—A law embodied in the equation $C = \dfrac{E}{R}$, where C = current in amperes; E = E.M.F. in volts; and R = resistance in ohms. By this equation, if any two quantities be given, the third may be found.

OVERLOAD RELEASE.—An attachment to a motor starting switch that causes it to open the circuit when the motor is subjected to heavy overload.

PILOT VOLTMETER.—A voltmeter connected to the far end of a feeder by special wires, with the object of ascertaining the pressure of supply at that point.

PLOUGH.—The current collector which runs in the road slot of a conduit tramway system. Also applied to certain switch operating apparatus used on surface contact tramways.

POLYPHASE CURRENT.—Two or three phase as opposed to single phase.

REGENERATIVE SYSTEM.—A system of electric traction in which the motors operate as dynamos when running down hill or stopping, thus returning power to the line.

RESISTANCE.—The opposition afforded by all conductors to the flow of electricity.

ROTOR.—The rotory element of an alternating current induction motor.

SECTIONALISED.—A street cable is sectionalised when it is split up into sections electrically connected together in special boxes by links.

SECONDARY CELL.—See Accumulator.

SERIES.—The connection of electrical apparatus or conductors in such a manner that the same current flows through all in succession.

SERIES-PARALLEL CONTROL.—An economic method of speed control where two or more motors are employed on a car. The motors are first switched in series and then in parallel.

SERIES WINDING.—The magnet winding of a series wound dynamo; i.e., one in which the whole of the current delivered by the armature passes through the field circuit. The same winding is used for motors.

SHOES.—The current collectors which make contact with the live rail of a railway system.

SHUNT.—A branch circuit which shunts part of the current away from the main circuit to which it is connected.

SHUNT WINDING.—The magnet winding of a shunt wound dynamo; i.e., one in which only about two per cent. of the full armature current is passed round the field circuit formed of fine wire. The same winding is used for motors.

SHORT CIRCUIT.—The provision of an alternative path for the current of much lower resistance than that of the main circuit.

SINGLE PHASE ALTERNATING CURRENT.—A simple alternating current necessitating a circuit of only two wires.

SKATE.—The current collector of a surface contact tramway system.

STATOR.—The stationary element of an alternating current induction motor.

SUB-STATION.—A building, subsidiary to the main generating station, where transforming operations are carried out.

SURFACE CONTACT.—A system of electric traction involving contacts, fixed level with the road surface, from which current is collected.

SYNCHRONISING.—The act of connecting an alternating current generator in parallel with another at the right moment of time.

SYNCHRONOUS MOTOR.—An alternating current motor which runs in exact synchronism with the generator from which the current is derived.

THERMO PILE.—An electrical generator, by means of which heat energy is directly converted into electrical energy. Of little practical use, and only capable of evolving a low E.M.F.

THREE PHASE CURRENT.—Three alternating currents generated simultaneously by one machine in separate windings and differing from one another in phase by a fixed amount.

THREE WIRE SYSTEM.— A system of distributing electrical power.

THROW-OVER SWITCH.—A switch provided with two or more contacts by means of which current is directed from one circuit to another.

TORQUE.—Turning effort.

TRANSFORMER.—An apparatus for converting electrical energy from one form to another. Generally used for pressure conversion.

TRIPLE CONCENTRIC CABLE.—A cable containing three concentric conductors insulated from one another.

TURNBUCKLE.—A screw arrangement whereby the tension of a span wire is adjusted.

TWO PHASE CURRENT.—Two alternating currents generated simultaneously by one machine in separate windings and differing from one another in phase by a fixed amount.

UNIT (BOARD OF TRADE).—The measure by which electricity is sold. It equals 1,000 watt-hours.

VARIABLE RATIO TRANSFORMER.—An alternating current transformer designed to yield a variable secondary voltage by alteration of the number of turns in circuit in the secondary winding.

VOLT.—The unit of electrical pressure or electro-motive-force.

WATT.—The unit of electrical power. 1 volt × 1 ampere = 1 watt.

INDEX.

BRADBURY, AGNEW, & CO. LD., PRINTERS, LONDON AND TONBRIDGE.

A SELECT LIST

OF

STANDARD WORKS

PUBLISHED BY

ARCHIBALD CONSTABLE & CO. LTD.

INDEX.

Electrical Engineering.

Electric Railway Engineering. By H. F. PARSHALL, M.Inst.C.E., etc., and H. M. HOBART, M.I.E.E. 475 pages and nearly 600 Diagrams and Tables. *Imp. 8vo.* 42s. *net.*

Undoubtedly the most comprehensive work on this immensely important branch of engineering, and likely to remain for long the standard book on the subject.

Electrical Engineer.—" Thoroughly up-to-date, and deals with the most recent practice. We can heartily commend it to all who are interested in the development of electric traction, although the names of the Authors form perhaps the best commendation."

Electric Railways: Theoretically and Practically Treated. By SIDNEY W. ASHE, B.S., Late Instructor in Electric Traction for the Brooklyn Rapid Transit Co., and J. D. KEILEY, Assistant Electrical Engineer, N.Y.C. and H.R.R.R. With numerous Diagrams. *Demy 8vo.* 10s. 6d. *net.*

A work for the student, the professional man, and the manufacturer.

Electrician —" We can commend it with great confidence to all engineers who have to do with the designing of electric railway rolling stock, either in regard to particular lines or as manufacturers."

Fourth Edition, Revised and Enlarged.

Electric Power Transmission. A Practical Treatise for Practical Men. By LOUIS BELL, Ph.D., M.Am.I.E.E. With 350 Diagrams and Illustrations. *Demy 8vo.* 16s. *net.*

Experimental Electro-Chemistry. By N. MONROE HOPKINS, Ph.D., Assistant Professor of Chemistry in The George Washington University, Washington, D.C. With 130 Illustrations. *Demy 8vo.* 12s. *net.*

Second Edition, Revised.

Practical Electro-Chemistry. By BERTRAM BLOUNT, F.S.C., F.C.S., Assoc.Inst.C.E. Illustrated. *Demy 8vo.* 15s. *net.*

The utility of this work is proved by the reputation it has attained since its publication.

Second Edition.

Electric Furnaces. By J. WRIGHT. With 57 Illustrations. *Demy 8vo.* 8s. 6d. *net.*

Electricity. —"To all who desire full information regarding| electrical furnaces this book should be the most useful."

Motors.

Motor Vehicles and Motors: Their Design, Construction and Working, by Steam, Oil, and Electricity. By W. WORBY BEAUMONT, M.Inst.C.E., M.Inst.M.E., M.Inst.E.E. Two volumes. *Imp. 8vo.* Vol. I. Second Edition, Revised. 636 pages and more than 450 Illustrations and Working Drawings. 42s. *net.* Vol. II. Nearly 700 pages and over 500 Illustrations. 42s. *net.*

The Autocar.—" The author is to be congratulated on having produced a work which not only completely overshadows anything which has been previously done in this direction, but which is also unlikely to be even nearly approached for some time to come. . . . An all-round book dealing with the technical aspects of automobilism."

New Edition, 1907.

O'Gorman's Motor Pocket Book. By MERVYN O'GORMAN, M.Inst.C.E., M.Inst.Mech.E., etc. *Limp Leather Binding.* 7s. 6d. *net.*

This book is a guide to those who wish to learn how to buy, drive, maintain, repair, tour with, and keep up-to-date a modern motor car. To do this, it explains in simple words the purpose of the large number of parts which together make up a car, and which are liable to abuse if not understood. The book is in dictionary form, with French, German, and Italian equivalents, each part, accessory, or tool having a brief article to itself, as well as diagrams in 301 instances.

Gas, Gasoline, and Oil Engines.

Gas Engine Design. By CHARLES EDWARD LUCKE, Ph.D., Mechanical Eng. Dept., Columbia University, New York City. Numerous Designs. *Demy 8vo.* 12s. 6d. *net.*

Gas and Oil Engine Record.—" Can be thoroughly recommended as a work of reference and as a storehouse of reliable information, not hitherto obtainable in a single volume.

Second Edition.

Gas Engine Construction. By HENRY V. A. PARSELL, Jr., M.A.I.E.E., and ARTHUR J. WEED, M.E. Fully Illustrated. 304 pages. 10s. 6d. *net.*

Fifteenth Edition, Reset and Enlarged.

Gas, Gasoline, and Oil Engines, including Gas Producer Plants. By GARDNER D. HISCOX, M.E., Author of " Mechanical Movements," " Compressed Air," etc. Fully Illustrated. *Demy 8vo.* 10s. 6d. *net.*

Power Plants, Machinery, Cranes, etc.

Entropy: or, Thermodynamics from an Engineer's Standpoint, and the Reversibility of Thermodynamics. By James Swinburne, M.Inst.C.E., M.I.E.E., etc. Illustrated with diagrams. 4s. 6d. net.

The Economic and Commercial Theory of Heat Power Plants. By Robert H. Smith, A.M.I.C.E., M.I.M.E., M.I.E.E., etc., Prof. Em. of Engineering. Numerous Diagrams. *Demy 8vo.* 24s. net.

Engineering.—"The book is a monument of painstaking investigation, and the collection of useful data it contains must prove of great value to those concerned in the planning of power plants.

Machine Design. By Charles H. Benjamin, Professor of Mechanical Engineering in the Case School of Applied Science. Numerous Diagrams and Tables. *Demy 8vo.* 8s. net.

Prepared primarily as a text-book, but containing mainly what the writer has found necessary in his own practice as an engineer.

Tenth Edition.

Mechanical Movements, Powers, Devices and Appliances. By Gardner D. Hiscox, M.E., Author of "Gas, Gasoline, and Oil Engines, etc. Over 400 pages. 1646 Illustrations and Descriptive Text. *Demy 8vo.* 12s. 6d. net.

Mechanical Appliances. Supplementary Volume to Mechanical Movements. By Gardner D. Hiscox, M.E. 400 pages. About 1,000 Illustrations. 12s. 6d. net.

Fourth Edition.

Compressed Air: Its Production, Uses, and Applications. By Gardner D. Hiscox, M.E., Author of "Mechanical Movements, Powers, Devices," etc. 820 pages. 545 Illustrations. *Demy 8vo.* 20s. net.

Cams, and the Principles of their Construction. By George Jepson, Instructor in Mechanical Drawing in the Massachusetts Normal School. 6s. net.

Power Plants, Machinery, etc.—*(continued)*.

Hydraulics, Text-Book of: Including an Outline of the Theory of Turbines. By L. M. HOSKINS, Professor of Applied Mathematics in the Leland Stanford Junior University. Numerous Tables. Fully Illustrated. *Demy 8vo.* 10s. 6d. net.

Forms a comprehensive text-book, but is more compact than the ordinary treatise on the subject.

Hydraulic Machinery. By A. H. Gibson.

Cranes. By ANTON BÖTTCHER. Translated from the German. Enlarged, and edited with a Complete Description of English and American Practice, by A. TOLHAUSEN, C.E. Fully Illustrated. 42s. *net.*

This volume deals with and fully explains and describes all the most recent machinery for the lifting and handling of work, both indoors and out—light and heavy.

Hardening, Tempering, Annealing and Forging of Steel. A Treatise on the Practical Treatment and Working of High and Low Grade Steel. By JOSEPH V. WOODWORTH. 288 pages. With 201 Illustrations. *Demy 8vo.* 10s. *net.*

Dies: Their Construction and Use for the Modern Working of Sheet Metals. By JOSEPH V. WOODWORTH. 384 pages. With 505 Illustrations. *Demy 8vo.* 12s. 6d. *net.*

Modern American Machine Tools. By C. H. BENJAMIN, Professor of Mechanical Engineering Case School of Applied Science, Cleveland, Ohio, U.S.A., Member of American Society of Mechanical Engineers. With 134 Illustrations. *Demy 8vo.* 18s. *net.*

Precision Grinding. A Practical Book on the Use of Grinding Machinery for Practical Machine Men. By H. DARBYSHIRE. Pages, viii. + 162. With illustrations. *Price 6s. net.*

Increased knowledge of this craft will, on practical and economic grounds, lead to a very general adoption of grinding processes in workshops. Owing to the exquisite degree of accuracy obtainable, it is likely to supplant entirely the older methods of finishing detail parts.

Fourth Edition.

Shop Kinks. A Book for Engineers and Machinists Showing Special Ways of doing Better, Cheap and Rapid Work. By ROBERT GRIMSHAW. With 222 Illustrations. 10s. 6d. *net.*

The book is indispensable to every machinist and practical engineer.

Construction, etc.

Third Edition, Revised and Enlarged, of this Standard Work.

Reinforced Concrete. By CHARLES F. MARSH, M.Inst.C.E., A.M.Inst.M.E., M.Amer.Soc.C.E., and W. DUNN, F.R.I.B.A. 654 pages and 618 Illustrations and Diagrams. 31s. 6d. *net.*

Engineer.—"Will be found useful as a book of reference . . . a guide to the student, and the builder and constructor will find it an extensive repertory of existing examples which he may consult with advantage."

Building News.—"It is worth most careful study; it will well repay the architect or engineer desirous of information about a structural combination the full possibilities of which probably few of us have yet grasped."

Second Edition, Revised.

Reinforced Concrete Construction. By A. W. BUEL and C. S. HILL. Fully Illustrated. 21s. *net.*

Economic Use and Properties of Reinforced Concrete. Beams and Theories of Flexure. Columns. Retaining Walls, Dams, Tanks, Conduits and Chimneys. Tests and Designs of Arches. Foundation Construction. Building Construction. Bridge and Culvert Construction. Conduit Construction. Tank and Reservoir Construction. Materials Employed. Various Methods Employed in Concrete Construction.

Concrete Blocks: Their Manufacture and Use in Building Construction. By H. H. RICE and WM. M. TORRANCE. Illustrated. *Demy 8vo.* 8s. *net.*

The Builder.—"Addressed to those who contemplate the manufacture and use of concrete blocks. Full of really practical information, contains numerous illustrations that are likely to be of great service to the architect and builder."

Second Edition, Revised.

Cement and Concrete. By LOUIS CARLTON SABIN, B.S., C.E., Assistant Engineer, Engineer Department, U.S. Army; Member of the American Society of Civil Engineers. *Large Demy 8vo.* 21s. *net.*

Surveyor.—"Mr. Sabin's notable book is as comprehensive as it is original in its investigations, and clear and concise in its arrangement."

Construction—*(continued)*.

The Elastic Arch: With Special Reference to the Reinforced Concrete Arch. By BURTON R. LEFFLER, Engineer of Bridges, Lake Shore and Michigan Southern Railway. Diagrams and Tables. *Crown 8vo. Price 4s. net.*

The writer is the first to present a correct and simple method of designing a reinforced concrete section, for combined thrust and moment.

Tunnel Shields and the Use of Compressed Air in Sub-aqueous Works. By WILLIAM CHARLES COPPERTHWAITE, M.Inst.C.E. With 260 Illustrations and Diagrams. *Crown 4to. 31s. 6d. net.*

The Engineer.—"This standard work . . . this valuable acquisition to engineering literature."

Modern Tunnel Practice. By DAVID McNEELY STAUFFER, M.Am.S.C.E., M Inst.C.E. With 138 Illustrations. *Demy 8vo. 21s. net.*

CONTENTS. Tunnel Location and Surveying. Explosives. Blasting. Notes on Shaft-Sinking. Principles of Tunnel Timbering and Driving. Tunnel Arch Centres. Sub-aqueous Tunnels and Tunnel Shields. Subway or Underground Railways. Special Tunnel-building Plant. Some Data upon the Cost of Tunnelling. The Ventilation of Tunnels. Air Locks. Tunnel Notes.

Earthwork and its Cost. By H. P. GILLETTE, Assoc. Ed., *Engineering News*, late Assistant, New York State Engineer. 256 Pages. Fully Illustrated. *8s. net.*

Second Edition, Enlarged.

Steel Mill Buildings: Their Design and the Calculation of Stresses in Framed Structures. By M. S. KETCHUM, Professor in Civil Engineering, University of Colorado. 380 Pages. Fully Illustrated. *16s. net.*

Bridge and Structural Design: With Tables and Diagrams. By W. CHASE THOMSON, M.Can.Soc.C.E., Assist. Engineer. Dominion Bridge Co. *Demy 8vo. 8s. net.*

Refuse Disposal and Power Production. By W. F. GOOD-RICH. *Demy 8vo.* Fully Illustrated. *21s. net.*

Small Dust Destructors for Institutional and Trade Refuse. By W. F. GOODRICH. *Demy 8vo. 4s. net*

Technological and other Works.

Agglutinants and Adhesives of all Kinds for all Purposes. By H. C. STANDAGE. *Demy 8vo. 6s. net.*

By following the order specified in each recipe it is possible to mix what otherwise appear to be incompatible bodies, such for instance as turpentine, paraffin, or benzine, with water, etc.

Glues and Gelatine: A Practical Treatise on the Methods of Testing and Use. By R. LIVINGSTON FERNBACH. *Demy 8vo. 10s. 6d. net.*

A work of the greatest economic value to consumers.

Cotton. By Prof. C. W. BURKETT and CLARENCE H. POE. Fully Illustrated. *Demy 8vo. 8s. 6d. net.*

This important work, written by eminent authorities, one of them being Professor of Agriculture in the North Carolina College of Agriculture, furnishes a very complete history of Cotton from earliest times, and a mass of practical information on its structure, botanical relations, the varieties of cotton and their classification, details of the world's supply, and the methods of raising and improving the plant. Further, it deals fully with the marketing, prices, manufacture and by-products.

Cotton Seed Products. A Manual of the Treatment of Cotton Seed for its Products and their Utilization in the Arts. By LEEBERT LLOYD LAMBORN, Member of the American Chemical Society; Member of the Society of Chemical Industry. With 79 Illustrations and a Map. *12s. 6d. net.*

Searchlights: Their Theory, Construction and Application. By F. NERZ. Translated from the German by CHARLES RODGERS. Very fully Illustrated. *7s. 6d. net.*

Aerodynamics. By F. W. LANCHESTER. Fully Illustrated. *Demy 8vo. 21s. net.*

Third Edition.

Enamelling. On the Theory and Practice of Art Enamelling upon Metals. By HENRY CUNYNGHAME, M.A., F.R.S. Two Coloured Plates and 20 Illustrations. *Crown 8vo. 6s. net.*

Physics, Chemistry, Astronomy, etc.

Principles of Microscopy. Being an Introduction to Work with the Microscope. By Sir A. E. WRIGHT, M.D., F.R.S., D.Sc., Dublin (Honoris Causa) F.R.C.S.I. (Hon.). With many Illustrations and Coloured Plates. *Imp.* 8vo. 21s. *net*.

LIST OF CONTENTS: General Considerations with regard to the Object Picture. Development of the Object Picture in the case where Objects are disposed upon a Single Optical Plane. Development of the Object Picture in the case where the Elements, which are viewed by the Unaided Eye, or, as the case may be, by the Microscope, are piled one upon the other. On the principles with govern the Staining Operations which are undertaken for the purpose of developing a Microscopic Colour-Picture. On the restrictions which hamper the Microscopist in the matter of the Selection and Development of the Stage-Picture; on the Defects and Sources of Fallacy which attach to the Microscopic Outline-Picture and Microscopic Colour-Picture respectively; and on the limitations which are imposed upon Microscopic achievement by the inadequate representation of the object in the Stage-Picture. Image Formation by the Simple Aperture On Image Formation by the Lens-armed Aperture. On the defects which are introduced into the Image by Spherical and Chromatic Aberration in the Lens. On the defects introduced into the Image by Diffraction occurring in the Aperture of the Lens. Image Formation in the case where the Object is viewed by the Unaided Eye. Image Formation in the case where an Object is viewed through a Simple Microscope. Image Formation in the case where the Object is viewed through the Compound Microscope. On the Optical Elements of the Microscope considered separately. On the Instrumental Adjustments required for the achievement of Dark Ground Illumination and Stereoscopic Illumination in particular in connection with Wide-angled High-power Objectives. On the Adjustments required for the achievement of a Critical Image. On the Question as to whether there are Definite Limits imposed upon Microscopic Vision and Resolution; and on the Real Nature of the Limitations encountered in connection with the employment of High Magnifications.

Radio-Active Transformations. By ERNEST RUTHERFORD, F.R.S., Professor of Physics at the McGill University, Montreal, Canada. Fully Illustrated. *Demy* 8vo. 16s. *net*.

Professor Rutherford's investigations into the properties of radio-active matter and his many discoveries have placed him in the front rank amongst the world's physicists. This book embodies the results of his researches in the field of radio-active bodies, where he has long been recognised as a leader among modern scientists.

Physics, Chemistry, etc.—*(continued)*.

Electricity and Matter. By J. J. THOMSON, D.Sc., LL.D., Ph.D., F.R.S. *Price 5s. net.*

Representation of the Electric Field by Lines of Force. Electrical and Bound Mass. Effects due to the Acceleration of Faraday Tubes. The Atomic Structure of Electricity. The Constitution of the Atom. Radio-activity and Radio-active Substances.

The Discharge of Electricity through Gases. By J. J. THOMSON. *Crown 8vo. 4s. 6d. net.*

The Electrical Nature of Matter and Radio-activity. By H. C. JONES, Professor of Physical Chemistry in the Johns Hopkins University. *8s. net.*

Experimental and Theoretical Applications of Thermodynamics to Chemistry. By PROFESSOR WALTHER NERNST, University of Berlin. *Extra Crown 8vo. 5s. net.*

Second Edition.

Liquid Air and the Liquefaction of Gases. By T. O'CONOR SLOANE, M.A., M.E., Ph.D. Many Illustrations. *10s. 6d. net.*

Physics. Heat. Heat and Gases. Physics and Chemistry of Air. The Royal Institution of England. Michael Faraday. Early Experimenters and their Methods. Raoul Pictet. Louis Paul Cailletet. Sigmund von Wroblewski and Karl Olszewski. James Dewar. Chas. E. Tripler. The Joule-Thompson Effect. The Linde Apparatus. The Hampson Apparatus. Experiments with Liquid Air. Some of the Applications of Low Temperature.

Van Nostrand's Chemical Annual: A Handbook of Useful Data for Analytical, Manufacturing and Investigating Chemists and Chemical Students. Edited by JOHN C. OLSEN, M.A., Ph.D., Professor of Analytical Chemistry, Polytechnic Institute, Brooklyn ; formerly Fellow Johns Hopkins University ; Author of " Quantitative Chemical Analysis, by Gravimetric, Electrolytic, Volumetric and Gasometric Methods." With the co-operation of eminent Chemists. *Crown 8vo.* Nearly 100 Tables. *12s. 6d. net.*

The book is concerned chiefly with numerical data because other books have not aimed at covering this field. All tables and data are quoted from the original source wherever possible.

Physics, Chemistry, etc.—(continued).

Practical Methods of Inorganic Chemistry. By F. MOLLWO PERKIN, Ph.D., Head of Chemistry Department, Borough Polytechnic Institute, London. With Illustrations. 2s. 6d. net.

Laboratory Note Book for Chemical Students. By VIVIAN B. LEWES, Professor of Chemistry, Royal Naval College, and J. S. S. BRAME, Demonstrator in Chemistry, Royal Naval College, and Assistant Examiner in Chemistry, Science and Art Department. Interleaved with Writing Paper. 4s.

Second Edition

Modern Astronomy. Being some Account of the Revolution of the Last Quarter of the Century. By. PROFESSOR H. H. TURNER, F.R.S. Illustrated. *Crown 8vo.* 6s. net.

Time and Clocks: A Description of Ancient and Modern Methods of Measuring Time. By H. H. CUNYNGHAME, M.A., F.R.S. Illustrated. *Crown 8vo.* 6s. net.

Affords a great deal of interesting information concerning all the known methods employed in various ages for measuring and indicating time : and by the aid of many practical illustrations lucidly and popularly explains the principle of the sundial, the water clock, the portable sundial (forerunner of the watch), the grandfather's clock, the chronometer, and other ancient and modern time measurers, as well as the application of electricity as a means of impulse in clocks.

The Seven Follies of Science. A Popular Account of the Most Famous Scientific Impossibilities, and the Attempts which have been made to solve them. To which is added a Small Budget of Interesting Paradoxes, Illusions, and Marvels. By JOHN PHIN, Author of " How to Use the Microscope "; " The Workshop Companion "; " The Shakespeare Cyclopædia "; Editor Marquis of Worcester's " Century of Inventions "; etc. With numerous Illustrations. *Demy 8vo.* 5s. net.

Lightning Source UK Ltd.
Milton Keynes UK
UKHW03f0200080618
323880UK00006B/943/P